Above: For the first trial model a 7.5 cm Pak 40 was mounted on a relatively high pedestal to clear the sheet-metal crew cab and enable all-round traverse of the gun. (HLD) Below: Firing trials were conducted with this first Versuchs 7.5 cm Pak40/4 auf RSO on 13 July 1943. (HLD)

Above, Below, and Upper Right: Still using the same Versuchs-RSO (Fgst.Nr. V4) chassis, Steyr had mounted an "improved" 7.5 cm Pak40/4 with a lower, dismountable pedestal and had fabricated a lower sheet-metal crew cab by 30 July 1943. (HLD)

Introduction

This last volume of the Panzer Tracts 7 series covers eight different self-propelled guns created for use by **Panzerjaeger** (tank destroyer) units. There is no common thread between these various design projects. Some were mounted on existing chassis (such as the **7.5 cm Pak 40/4 auf RSO** and the **7.5 cm Pak 44 auf 3 to Zgkw.**). Others were created with newly designed chassis, but still utilized already existing automotive components (such as the **Panzerjaeger Hornisse, Waffentraeger**, and the **Pz.Sfl. fuer 12.8 cm Kan.40**). However, they all had two features in common - open tops so that the crews could observe and hear much better than tank crews and light armor to keep the weight down for better mobility.

In all cases these **Selbstfahrlafetten** (self-propelled carriages) were more effective weapon systems for knocking out tanks than the same gun on a towed carriage. As reported by the user organizations, high velocity anti-tank guns weighing more than a ton were too heavy for the crews to move to alternative firing positions and were too big to hide in front line positions with the infantry. But, as proven by their kill-to-loss records, these same heavy guns mounted on self-propelled chassis could be very effectively used in counterattacks to strike the flank of enemy tank attacks and also could rapidly change positions when they themselves came under fire.

From 1943 to the end of the war, **Wa Pruef 4** (artillery design office) kept insisting that these **Selbstfahrlafetten** have two features: 360 degree traverse for all-round fire and the gun be dismountable for emplacement on the ground. While the first requirement was achieved with the **7.5 cm Pak 40/4 auf RSO** and **8.8 cm Pak 43 auf Waffentraeger**, the second requirement to dismount the gun onto a towed carriage kept getting in the way of completing projects for mass production. This requirement to dismount the gun is not understood. The main advantage appears to be that the gun could still be kept in action when the vehicle broke down. But, couldn't the same objective have been met simply by placing the **7.5 cm Pak 40** with its original wheeled carriage on an unmodified **RSO** (just like the British mounted their 2 and 6 pounder anti-tank guns "portee" on trucks)?

But "simple" solutions were not acceptable to German engineers. Fortunately some of the correspondence and records of meetings between **Wa Pruef** (ordnance design office) and engineers from the detail design firms have survived to provide us with some insight into the basis for their design decisions. All reports that have been found were translated and included in the text to provide clues to the circumstances forcing them to take certain design directions. Even so, our inability to now question these key individuals leaves us with some unresolved contradictions. For example, why did they abandon the single axle carriage for the **8.8 cm Pak 43/41** in favor of the heavier dual axle carriage for the **8.8 cm Pak 43** - and then state that a single axle carriage was needed for the dismounted **7.5 cm Pak 42** because *there was a shortage of heavy towing vehicles needed to tow a twin-axle carriage?*

A true understanding of the value of the **Panzerjaeger** can be gained from experience reports written by unit commanders close to the time when the action occurred. Based on their personal experience in combat, the observations and advice contained in these wartime reports totally contradict misinformed statements from self-proclaimed experts whose avid interest can hardly compensate for their lack of experience, lack of technical knowledge, and inability to conduct objective weapons system analyses. If the **12.8 cm Kanone 40** was such a superior long-range tank killer, why was it being used to knock out T34 tanks at 1000 to 1500 meters - ranges in which it could itself be penetrated by the 7.62 cm gun on the T34? If the **8.8 cm Pak 43** was so effective at killing tanks why did a company commander refuse to engage a battalion of T34 tanks? If the **7.5 cm Pak 40/4 auf RSO** was such a great idea, why did the units using it report: *Although the* **Pak 40 auf RSO** *can be immediately brought into action and is maneuverable, it has been a complete failure.*

Over one thousand hours were spent carefully measuring and drawing all the details on a surviving **Panzerjaeger "Hornisse"** and a **7.5 cm Pak 40/4 auf RSO** to create ultra-accurate as-built drawings. In addition, about a hundred "clean" photos taken during the war were carefully examined to determine precise details of bits now missing from the "survivors" and to determine the exact order in which modifications were introduced into the production series. It is this careful and thorough study of details, combined with over 37 years spent searching for all the surviving documents, drawings, and photos, that makes Panzer Tracts the ultimate in historically accurate documentation.

Measured and drawn at 1:1, the drawings are printed at the popular 1:35 scale. Even so, there is no loss in detail with this reduction in scale because the software now being used to create the printing plates is not a line-by-line printer. When we draw a circle, the software draws a circle using the center and radius. When we draw an arc, it draws an arc using the two ends and center point. This is vastly superior to the old technique of laying down pixels whenever the line printer crosses a drawn object, resulting in saw-toothed edges on curves and slanted lines. Compare the super-clean drawings in Panzer Tracts 5-3, 7-1, 7-2, 7-3, 9-3, and 15-2 with the older line printer technique used to print cad drawings in Panzer Tracts 1-1, 1-2, 5-1, 13, 20-1, and 20-2.

P.S. The crew cab for the **7.5 cm Pak 40/4 auf RSO** was made out of thin sheet metal - not armor.

7.5 cm Pak 40/4 auf RSO

Following initial project discussions with **Wa Pruef**, Rheinmetall completed a conceptual design drawing H Sk B 82838 on 28 May 1943 of a **7.5 cm Pak 40** mounted on a pedestal on top of a modified **RSO** chassis. Planning was well advanced when this project was discussed in Speer's conference with Hitler on 4/5 August 1943: *Hitler is pleased that **Wa Pruef** working together with Steyr has developed the **Pak 40 auf Raupenschlepper Ost**. Firing trials have started and troop trials are to be quickly organized. This also applies to the proposed **Schlittenanhaenger mit Pak 40** (towed sled).*

On 2 September 1943, **Wa Pruef 4** reported the basic design ideas for the **7.5 cm Pak 40 auf RSO** and project status, as follows:

*__Panzerabwehr__ (anti-tank) experience reports continuously complain that because of its weight the **schweren Panzerjaegerkanonen mot Z** can't be moved by the crew on the battlefield. Changing firing positions and moving short stretches can only be done with assistance from towing vehicles that must first be brought forward. In addition to the loss of time, changing positions under enemy fire usually results in the loss of men and equipment. Easy maneuverability in combat is a requirement that must be demanded for any towed **Pak** with limited traverse associated with its **Spreizlafette** (split trails). Based on frontline experience in the East, this requirement is met only by guns that weigh up to 1500 kg, i.e., **leichten** and **mittleren Pak**.*

*Based on these facts, **Wa Pruef 4** initiated studies to determine under which specifications and which methods it would be possible to make it easier to move the 7.5 cm **Pak 40**. **Wa Pruef 4** started with the idea of creating a mechanized **Pak** that can immediately fire from the vehicle and pull out of enemy fire more rapidly than is possible with the towed **Pak** weighing 1.5 tons.*

*The **Raupenschlepper Ost** serves as the towing vehicle for the **7.5 cm Pak 40**. This led to the idea of mounting the gun with upper carriage on the **RSO** and, if possible, achieve all-round fire and dismountability. Working together with Rheinmetall-Borsig (the development company for the **7.5 cm Pak 40**), a **Sockel** was designed to mount the **Pak 40** with its upper carriage on the **RSO**. At the same time the firm Steyr-Daimler-Puch A.G. (development company for the **RSO**) was pulled in to possibly use an unmodified **RSO** for this purpose.*

*The first trials, made to determine if it was possible to fire to the side from the **RSO**, were completely successful. The **Fuehrerhaus** was dropped so that the firing height could be lowered to 1800 mm from 2100 on the original design (which is now more than 200 mm lower than the firing height of the available **Pak-Sfl.** on **Pz.II** and **38(t)** chassis, which have a limited traverse and aren't loved by the troops because of their height). Otherwise nothing is changed on the **RSO** except the **Pritschenaufbau** was changed so that 30 rounds of ammunition could be stowed (which is the same amount that is carried when an **RSO** tows this **Pak**).*

*For the first time a successful trial has been completed with a gun mounted on a previously available chassis which has both all-round fire and can be dismounted. This design also has significant advantages for **Pak 40** production because the gun with upper carriage can be*

Above: At an early project stage in June 1943, Steyr conducted trials with an unmodified 7.5 cm Pak 40 mounted on Versuchs-RSO (trial RSO) with Fgst.Nr. V4. (HLD)

Below: By 9 September 1943, an Allwetterverdeck (all weather cover) canvas cover had been added to help disguise this Versuchs 7.5 cm Pak40/4 auf RSO (Fgst.Nr.V4) as a normal transport vehicle. (HLD)

7-173

used without modification and some of the components that are difficult to acquire (such as wheels with rubber tires, axles, brakes, and torsion bars) are no longer needed. Developed by Steyr, a light crane for dismounting the gun is carried on the vehicle.

An additional requirement from **Wa Pruef 4** was that, for firing in winter the **Pak 40** is to be mounted on a sled with the same **Sockel** and all-round fire. Rheinmetall developed a sled with **Pritsche** constructed to stow 28 rounds of ammunition.

As ordered by Hitler, troop trials are to be accelerated. A contract has been awarded for six **Versuchsgeraete** that are planned to be completed by the end of September. In early September, **Wa Pruef 4** is conducting basic firing and driving trials.

The significant advantages of a **7.5 cm Pak 40** mounted on an **RSO** over a towed **Pak** were listed in an attached table, as follows:

o The **7.5 cm Pak 40 auf RSO** has a crew of four (**Geschuetzfuehrer**, 2 **Mann**, 1 **Fahrer**), while the towed **Pak 40** has a crew of eight.

o When pulling into a firing position, the **7.5 cm Pak 40 auf RSO** is ready to fire with a fresh crew after releasing the travel lock and mounting the gun sight. The towed **7.5 cm Pak 40** is ready to fire with a tired crew after the gun is uncoupled from the **RSO**, the trails spread, ready ammunition is unloaded from the **Munitionsfahrzeug**, and the **RSO** is driven into cover.

o When changing firing positions, the **7.5 cm Pak 40 auf RSO** is immediately mobile, the travel lock can be secured and the gun sight dismounted while driving. The towed **7.5 cm Pak 40** is mobile only after the **Zugmaschine** has been retrieved from cover, driven to the site, ammunition loaded, gun hooked up, travel lock secured, and gun sight dismounted before driving.

o The **7.5 cm Pak 40 auf RSO** has a 360 degree traverse, while the towed **7.5 cm Pak 40** has a 65° traverse arc.

o For cross-country travel the **7.5 cm Pak 40 auf RSO** has a ground clearance of 550 mm compared to 350 mm for the towed **7.5 cm Pak 40**.

o The **7.5 cm Pak 40 auf RSO** is more mobile with lower ground pressure, can be turned within the vehicle's own length, can reverse for long stretches, climbs up to 70% slopes, and can cross any depth of snow. While a turning circle of 13.6 m is necessary for the towed **7.5 cm Pak 40** with danger of damaging the trails by a smaller turning circle, reversing is almost impossible, it can only climb up to 40% slopes, and gets stuck in snow over 40 cm deep.

o The **7.5 cm Pak 40 auf RSO** weighs less (5,400 kg compared to 6,300 kg for the towed **7.5 cm Pak 40**), and therefore the fuel consumption is lower.

o There were also significant savings in **7.5 cm Pak 40 auf RSO** production due to less components, and it could be produced in a shorter time.

On 9 September 1943, the **Reichsminister fuer Bewaffnung und Munition** reported: *When shown a photo, Hitler expressed high approval of the proposal to protect and camouflage the* **Pak 40 auf Raupenschlepper Ost** *with an* **Allwetterverdeck** *(all weather cover). Therefore plans are to be made for it to go into production.*

On 30 September 1943, General Thomale, **Chef des Stabes for Genealinspekteur der Panzertruppen** reported: ***Gen.Insp.d.Pz.Tr.*** *agrees with the immediate further development of the* **7.5 cm Pak 40 auf RSO** *(***Absetzbar*** and* ***Rundumfeuer***). It is requested that broad-based troop trials be conducted on the Eastern Front in* **Panzerjager-Abteilungen** *in* **Infanterie-Divisionen** *and that 50 be ordered for this purpose. This requires: 1) a dismounting device for each gun, and 2) the dismounted gun must be movable for short stretches. The basis for this request is:*
a. Firing positions can then be changed without remounting the gun
b. The gun can be quickly pulled out of enemy fire and then remounted under cover,
c. In case the **RSO** *is lost, the gun can be expediently towed by another* **RSO***.*

In addition, determine if it is possible to design a simple gun limber in the form of a single axle trailer.

On 1 October 1943, the **7.5 cm Pak 40 auf RSO** was demonstrated for Hitler. In the Speer conference it was recorded that: *Hitler is completely pleased with this equipment especially because of the extraordinary winter mobility, suitability for mobile combat, and establishing* **Schwerpunkt** *(strong points). Disconnected from the 50* **Versuchsstueck** *that are to be completed for large troop trials, preparations are to be made for mass production of up to 400 per month. Since the necessary* **Pak 40** *and* **RSO** *for towing them are already being delivered, additional* **Sockel** *production is an acceptable risk.*

On 7 October 1943, **OKH GenStdH/Org.Abt.** reported: *The 50* **Pak 40 auf RSO***, first available in October, are to be issued to* **Heeresgruppe Mitte***: 28 for* **Armee Pz.Jg.Abt.743***, 12 for the Ski-Jaeger-Brigade and 10 as equipment reserve.*

On 1 November 1943, **Wa J Rue (Wu G 6)** reported on the **Raupenschlepper Ost** card that: *Plans are to produce 60* **RSO mit 7.5 cm Pak 40** *in addition to the* **Vorratsfahrzeuge** *(supply vehicles). Delivery date hasn't been determined. Two* **Versuchsfahrzeuge** *have been completed.*

On 11 November 1943, Hauptmann Windemuth, **Stab des Panzeroffiziers beim Chef GenStdH**, reported:
Firing trial results from the previously completed model of the **7.5 cm Pak 40 auf RSO** *haven't been satisfactory. Design changes are necessary. Development is very restricted because of continuous demonstrations.*

Two **RSO** *are needed for each* **Pak 40 auf RSO***;*

one for the gun and the other for the equipment (crane, towing device, etc,), ammunition and part of the crew that can't ride on the first **RSO**. It should be possible to achieve all of this with one **RSO**. Then the **Pak** would have to be towed in winter on a sled and in summer on a small trailer while the equipment, ammunition and crew are carried on the **RSO**. The advantage of using two **RSO** is that if one vehicle is lost, the **Pak** can be dismounted from the **RSO** or trailer using the **Sockel** with four poles.

In the meeting on production development on 21 January 1944: *Oberst Woehlermann (**Wa Pruef**) reported that development and the first Versuchsserie of 50 had been completed with the gun dismounted on a Kreuzlafette. Further production is to be made using an le.F.H. carriage. The Bockkran (jib crane) is the bottleneck. Approval for vehicle requisition and Bockkran is requested.*

At the Speer conference on 25 January: *Hitler was very interested in the reported results of the driving trials of the 7.5 cm Pak 40/4 auf RSO and awaits the scheduled production reported in the Programmplanung for 50 in March, 100 in April, 150 in May, 200 in June, and 400 each month starting in July 1944.* The **7.5 cm Pak 40/4 Sfl. RSO** was demonstrated for Hitler on 26 January 1944, and it was recorded in Speer's conference that: *The 50 completed 7.5 cm Pak 40/4 (Sf.) RSO are to be sent quickly for troop trials by Heeresgruppe Sued. Without waiting for experience reports, planned production of this weapon is to continue with a production goal of 400 per month until conversion to the leichte Sturmgeschuetz 38(t) is ordered.*

On 26 January 1944, **OKH GenStdH/Org.Abt.** reported that the 60 previously completed **7.5 cm Pak 40 auf RSO** have been ordered for troop trials in **Heeres Gruppe Mitte**.

On 18 March 1944, **A.O.K.18 Stopak** sent a telegram to the **OKH/GenStdH/Org.Abt.** informing them that: *The following experiences have been gained from employing the Pak 40 auf RSO:*
1. Employed as a self-propelled gun the advantages are an increase in the number of mobile anti-tank guns and it has a large traverse. Disadvantages are the high casualties because of the lack of a shell fragment shield for the crew.
2. Emplaced in defensive position it has the advantage of a large traverse. The disadvantages are that it is impossible to quickly change firing positions, impossible to change position, and the gun is lost when the vehicle is lost.
3. Proposed modifications include all-round fragment protection for the crew, mount the gun on a stronger vehicle, and mount wheels on the gun so that it can be towed expediently.

At the Speer conference on 22 May 1944: *Hitler decided that the Pak 40 auf RSO should be produced at the high rate ordered. In addition, the troops complaints against this current model were already recognized during demonstrations and at the time the decision was made to accept this weapon.*

On 4 June 1944, **OKH GenStdH/Org.Abt.** reported: *Cease production of the 7.5 cm Pak 40 auf RSO. As reported by the troops, although the Pak 40 auf RSO can be immediately brought into action and is maneuverable, it has been a complete failure. The main deficiencies are that it presents too large an unarmored target, is too slow, and the RSO is overloaded. Therefore, Hitler has decided that production of the 7.5 cm Pak 40 auf RSO will cease after all the previously contracted number have been delivered.*

Other than the first 60 in the **Versuchsserie**, **Wa J Rue (Wu G)** reports did not record that any additional **7.5 cm Pak 40/4 auf RSO** were produced

Operational Service

Units that reported having **7.5 cm Pak 40/4 auf RSO** include (in chronological order of first report):

Pz.Zerst.Batl.478 under **VI.SS Freiw.A.K.** with **10.Lw. Feld.Div.**
1Mar44 - 18 operational **7.5 cm Pak auf RSO**. Employment of the **RSO Sfl.** requires the highest mobility of the **Kompanie-Chef** and **Zugfuehrer** or the entire action is threatened with failure as was often demonstrated in the last few weeks. The **Bataillon** is 60% operational with three **Panzerjaeger-Kompanien** outfitted with **RSO/Sfl**.
1Apr44 - 16 operational, 7 repairable. The **Bataillon** is 50% operational with three companies
1May44 - 11 operational, 1 long repair. As a result of continuous losses, the **Bataillon** is 60% operational with two **Panzerjaeger-Kompanien** outfitted with **RSO/Sfl**.
1Jun44 - No report

Pz.Zerst.Batl.477 under **A.O.K.18**
1Apr44 had 8 **7.5 cm Pak RSO** which can't be used because the sights were lost during a bombing attack on the way forward.
1May44 - No **7.5 cm Pak RSO** reported

Pz.Jg.Abt.263 in the **263.Inf.Div.** under **A.O.K.16**
1May44 - 0 **Pak Sf**. As ordered by **A.O.K.16** on 27 April 1944, a **Beute Pz.Kp.** is attached that is to be reorganized as a **Pz.Jg.Sturmgesch. Kp.**
1Jun44 - 5 operational, 2 repairable **Pak Sf. RSO**. 7 **RSO Pak 40** have been temporarily issued to the **2.Pz.Jg.(Stu. Gesch.)Kp.**
1Jul44- 6 operational and 1 repairable **RSO**. The frames of all chassis on the **RSO Pak** have been deformed and in part broken.
1Aug44 - 4 operational and 3 repairable **Pak40 RSO**
1Sep44 - 2 operational and 1 repairable **RSO**. As a result of engine failure, 3 **RSO Pak** need long-term repairs. The **2.Kp** is being converted to **Sturmgeschuetz** in Mielau.

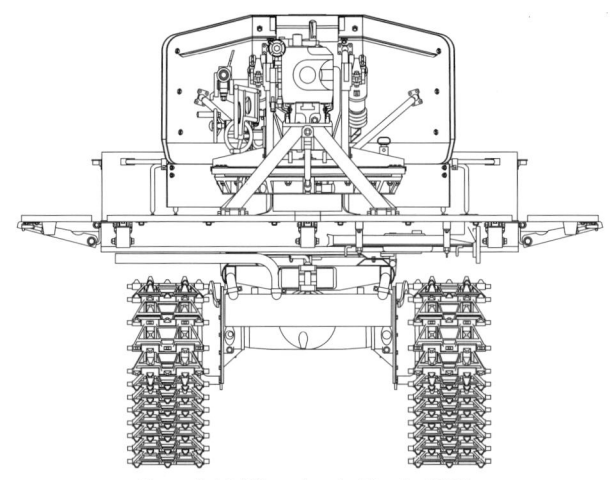
Copyright Hilary Louis Doyle 2006

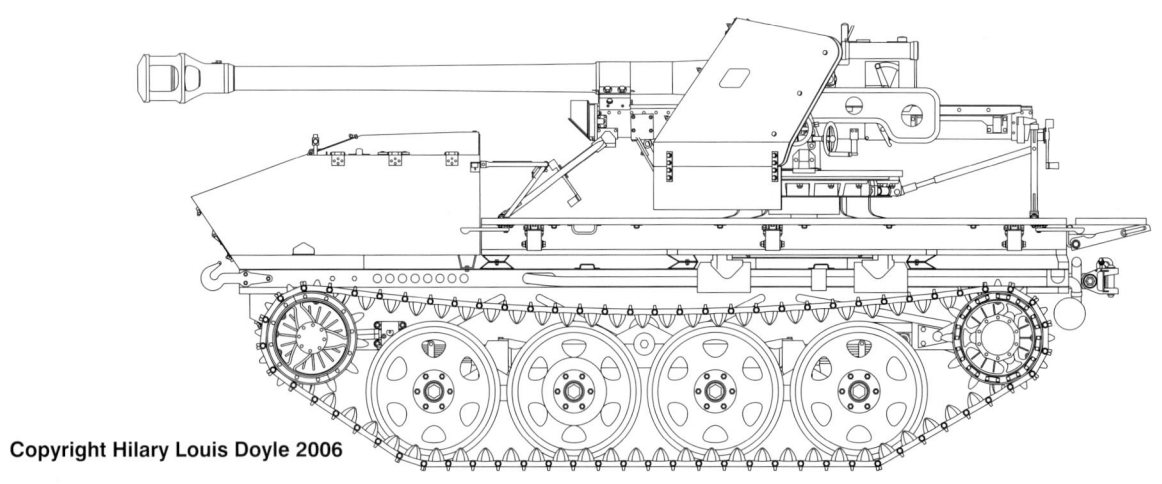
Copyright Hilary Louis Doyle 2006

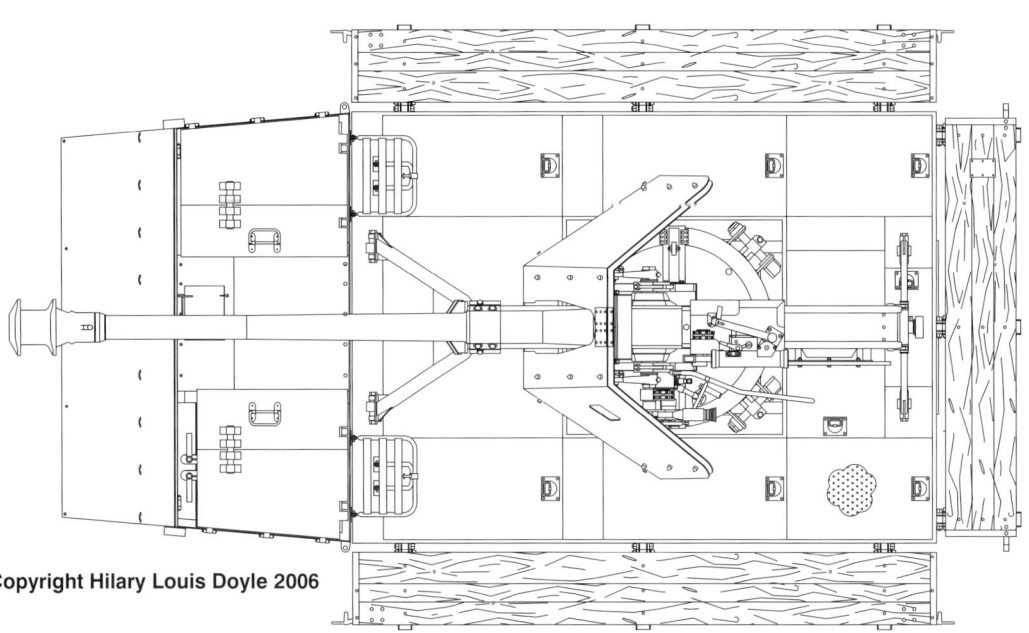
Copyright Hilary Louis Doyle 2006

7-176

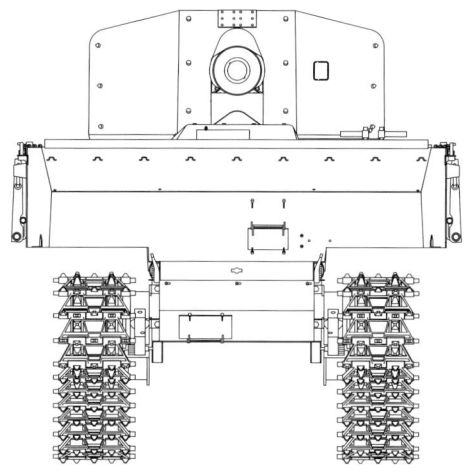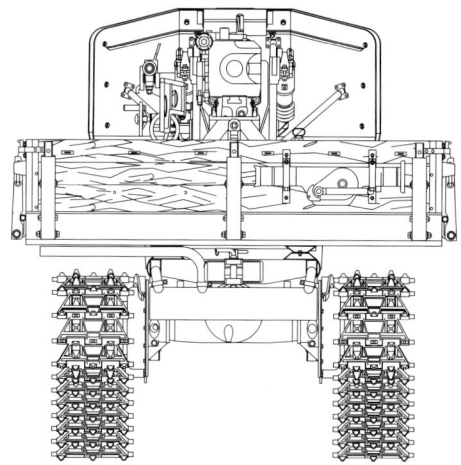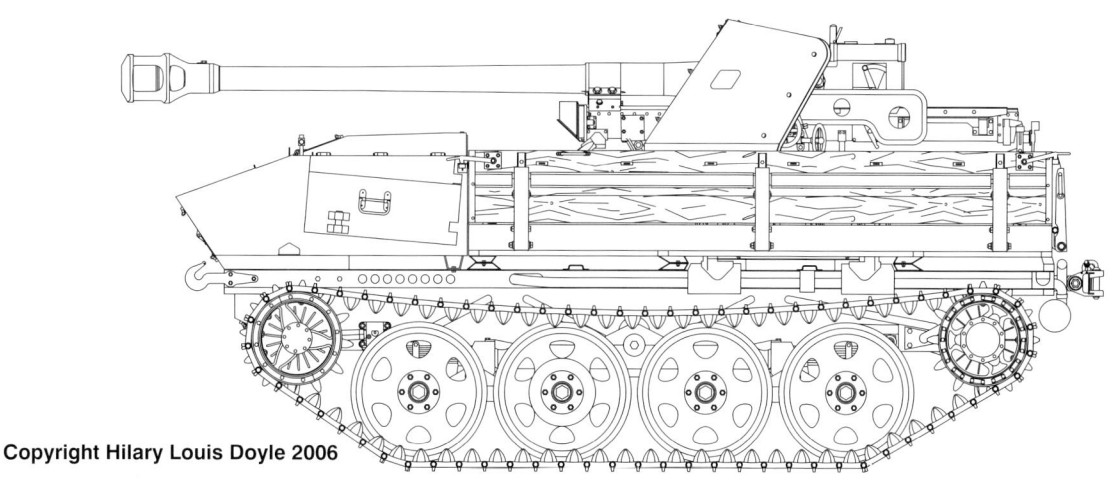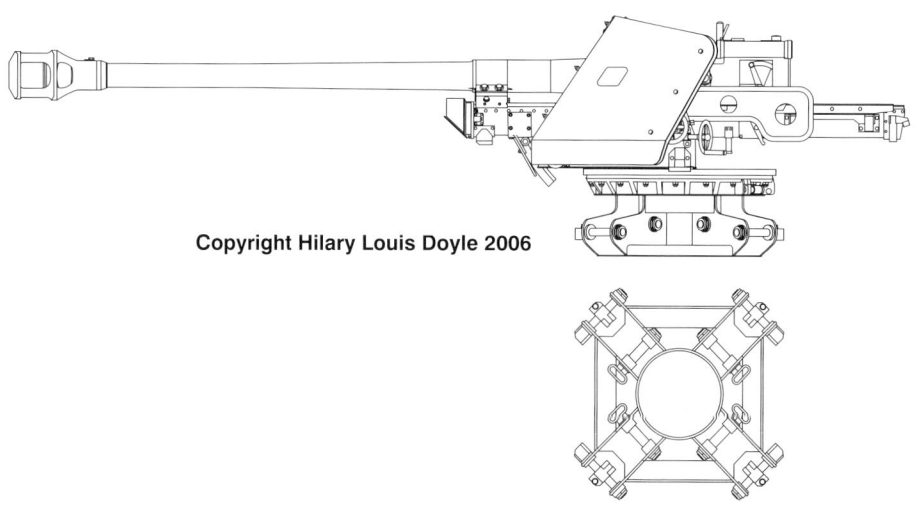

A Versuchsserie 7.5 cm Pak 40/4 auf RSO

Jan45 - 1 **Pak 40 RSO** available
Pz.Jg.Abt.752
1May44 - 7 operational, 1 repairable **Pak Sf. RSO**
1Jun44 - 6 operational, 2 repairable **Pak Sf. RSO**
1Jul44- **Pak 40 auf RSO** haven't worked out because they are too large a target and the **RSO** engines overheat in hot weather
205.Inf.Div. and 281.Sich.Div. - On 28 May 1944, **H.Gr. Nord** ordered that 3 **RSO Sf. 7.5 cm Pak 40** as a mobile platoon be issued to each of these units. With regard to its chassis, this weapon is not to be employed in or directly behind the **HKL** (main battle line).
Pz.Jg.Abt.751
1Sep44 - 3 **7.5 cm Pak auf RSO** present
1Oct44 - None

This Page:
This 7.5 cm Pak 40/4 auf RSO, used for Wa Pruef trials, was found by the Allies in Germany at the end of the War. A total of 28 rounds of 7.5 cm Pak 40 ammunition were stowed in seven bins with covers that formed most of the deck area. (NA)

Above: This Versuchsserie 7.5 cm Pak40/4 auf RSO (photographed at Steyr on 5 January 1944) had several minor modifications, including a deflector bar on the sheet-metal driver's cab. (HLD)
Below: A different 7.5 cm Pak 40/4 auf RSO was issued for troop training. (TTM)

7.5 cm Pak 40/4 auf RSO

Weapons Data: 7.5 cm Pak 40 (L/46)
   Elevation: -5, + 14 degrees
   Traverse: 360 degrees
   Gun Sight: Pak Z.F. 3 x 8 degrees
   Graduated to: 1800 Pzgr., 2800 Sprgr.

Secondary: None

Ammunition: 42 - 7.5 cm Pzgr. & Sprgr.

Crew: Commander, Gunner
Loader
Driver

Communication: No radio set

Measurements:
Length, overall: 4.71 m
Width, overall: 2.13 m
Height, overall: 2.10 m
Firing Height: 1.81 m
Wheel Base: 1.35 m
Track Contact: 2.07 m
Combat Loaded: 5.4 metric ton
Fuel Capacity: 180 liters

Armor Protection:
Gun Shield   2 x 4 mm
Driver's Cab   3 mm sheet metal (not armor)

Automotive Capabilities:
Maximum Speed: 17.2 km/hr
Range on Road: ?? km
Grade: 35 degrees
Trench Crossing: ?? m
Step: ?? m
Fording Depth: 85 cm
Ground Clearance: 55 cm
Ground Pressure: 0.38 kg/cm^2
Power Ratio: 13 HP/ton
Steering Ratio: 1.53

Automotive Components:
Motor: Steyr V-8, 3.5 Liter
   air-cooled
   3.517 liter
   70 HP @ 2500 rpm
Transmission: 4 forward, 1 reverse
Steering: Brakes/Differential
Drive: Rear sprocket
Roadwheels: 4 per side
Tires: Steel 660 mm dia.
Suspension: Leaf Springs
Track: 340 mm wide
   dry pin
Links per side: 69

Above:
Details of the rear travel lock and dismountable pedestal on the 7.5 cm Pak 40/4 auf RSO used for Wa Pruef trials.
(NA)

Below:
Details of the forward travel lock and extension on the gun shield on the 7.5 cm Pak 40/4 auf RSO.
(NA)

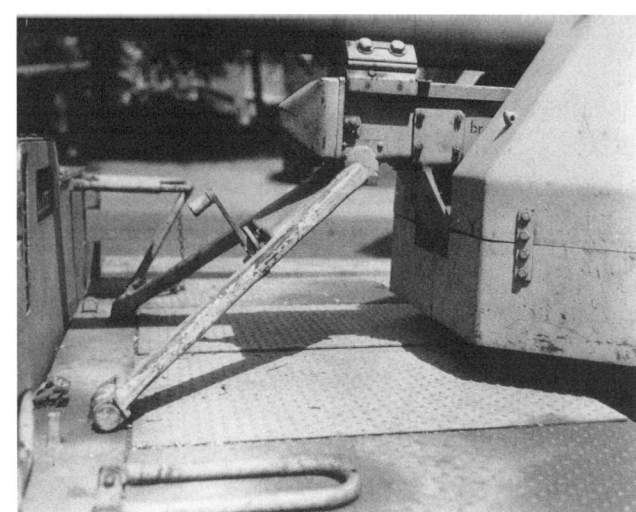

7.5 cm Pak 44 auf 3 to Zgkw.

The first mention in surviving documents of this project is in notes from the Speer conference with Hitler on 11 September 1943. During this meeting Hitler agreed with the proposed development of several **Waffenamt** self-propelled gun projects, including a **Pak 44 (7.5 cm L/70)** with **Rundumfeuer** (all-round fire) mounted on a **Zgkw. 3t** that was dismountable onto a **Kreuzlafette** (cruciform carriage).

However, early September could not have been anywhere near the starting date of the project because this **7.5 cm Pak 44 auf 3 to Zgkw.** had already been demonstrated to Hitler on 1 October 1943. As recorded in the Speer conference: *There are exceptional opinions in regard to whether the **7.5 cm Pak 44 auf 3-Tonnen Zugkraftwagen** has sufficient stability when the gun is fired to the side, and that in general the long gun is difficult to traverse and especially difficult to secure in a travel lock.*

*The ability to dismount this gun is to be based on the results of a troop trial. Due to its lack of mobility a **Kreuzlafette** shouldn't be considered. Due to the shortage of heavy towing vehicles, dismounting onto a twin-axle carriage can't be chosen. Therefore a single axle carriage must be rapidly developed and prepared for a thorough troop trial. In any case, the dismounting equipment is to be a simple **Bockkran** (jib boom) carried with the vehicle.*

After this demonstration on 3 October 1943, the **GenSt.d.H. Org.Abt.** reported: *The tactical requirement of all-round fire must be incorporated into the design of the **7.5 cm Pak 42 auf 3 to Zgkw.** If the design proves acceptable, a sufficient number should be produced to outfit an **Abteilung** each month (about 60 guns with replacements). Firing trial results are awaited before a final request is made.*

In a meeting between Hitler and the **General d. Artillerie** on 26 January 1944, a decision was made to cease developing the **7.5 cm Pak 42 (L/70) auf 3t Zgkw.**

A Waffentraeger Steyr (improved armored RSO with two engines for mounting a 7.5 cm StuK 40) that was demonstrated to Hitler on 26 January 1944.

"Hornisse" renamed "Nashorn"

Development

In mid-June 1942, as requested by Hitler, the **Waffenamt** embarked on a project to create an 8.8 cm anti-tank gun with the performance of the **8.8 cm Flak 41**, capable of penetrating 160 mm thick armor at 1000 m at 30 degrees. Both Rheinmetall (detailed design firm for the **8.8 cm Flak 41**) and Krupp (detailed design firm for the **8.8 cm Kw.K.41**) were given contracts to create different models.

On 23-25 July 1942 this project was discussed during Speer's conference with Hitler as items 33 and 34: *It will not be possible to produce the proposed 8.8 cm Pak 43 design with new components by the Spring of 1943. Therefore it should be quickly determined on which towed carriages and self-propelled carriages the new 8.8 cm Pak 43 gun can be mounted. 300 to 500 of these guns are to be delivered by 12 May. In addition, the three new designs currently in development are to be quickly completed.*

Because the le.F.H.43 auf Selbstfahrlafette (self-propelled gun) is an entirely new design, the three models are to be developed to the first Versuchsgeschuetz (trial guns). Both Rheinmetall's and Krupp's proposal to mount the 8.8 cm Pak 43 on the chassis of the le.F.H.43 (Sfl.) is to be pursued to a Versuchsstueck (trial piece).

Following a meeting on 28 July 1942, on 30 July, Dir. Spielvogel (Rheinmetall-Borsig A.G., **Waffenkonstruktion**) wrote to **Wa Pruef 4** for clarifications on the ammunition to be fired by the new **8.8 cm Pak 43**:

Because of the urgency of the work and to prevent repeating work, it is necessary to immediately decide on the which internal dimensions will be chosen for the gun. The Flak 41 Patrone (complete round with cartridge) has been tested and accepted and the lifespan of the gun tube determined by test firing.

We believe that the Flak 42 Patrone (from Krupp) is neither tested nor accepted. We heard that the Flak 42 Patrone is based on the 8.8 cm Kw.K. L/71. Because only a small number of these guns will be produced with first delivery at the end of the year, shouldn't the 8.8 cm Kw.K. L/71 be changed in the interest of standardizing cartridge production? Both Patronen are about the same length, the difference being in their diameter, with 117 mm for the 8.8 cm Flak 41 and 132 mm for the Flak 42. The fatter 8.8 cm Flak 42 cartridge allows a larger charge. But an increase in performance shouldn't be preferred when considering the decrease in the lifespan of the highly stressed gun tube.

Because of supply and production considerations, a new cartridge shouldn't be accepted. Also its larger diameter is unsuitable for the roller loader that has already been accepted by the Flak. L. Flak has decided that the Flak 42 will be mounted on the new self-propelled carriage. In the interest of standardizing ammunition, L. Flak will probably require that the Flak 41 Patrone be used in the Flak 42.

For standardization we have used the unaltered breech design from the 7.5 cm Pak 42 for the le.F.H.43 and the 8.8 cm Pak 43 in which either an electrical or mechanical firing mechanism can be used in the breech block. However, it is impossible to use this type of breech and breech block for the 8.8 cm Pak 43 if we are forced to use the 8.8 cm Flak 42 Patrone.

A decision was made by 5 September 1942 to replace the Rheinmetall cartridge with the larger diameter Krupp cartridge. For production reasons, the L/72 gun tube weighing about 1500 kg was to be constructed in two parts. From the breech to the tip of the muzzle, this **8.8 cm Pak 43** was slightly shorter (6280 mm) than a **8.8 cm Kw.K.43** (6298 mm) but much shorter than a **8.8 cm Flak 41**.

At a meeting at **Wa J Rue (WuG 2)** on 11 September 1942, the **Pak 43** was discussed by representatives from **Wa Pruef 4**, Krupp, and Rheinmetall, as follows: *It was determined that 500 Pak 43 must be produced by 12 May 1943. The gun tube is to be taken from the newly designed Pak 43. The carriage is from the 10 cm Kanone 41, which means complete new guns must be produced. The drawings for this new Pak 43/41 are to be completed by 31 October and a Versuchsgeschuetz fired by 20 December 1942. Raw material for the 500 carriages is to be provided by Rheinmetall, Duesseldorf. Production capacity for 50 guns per month has been secured at Henschel.*

Realizing that the new **Selbstfahrlafette** design projects (initiated in early 1942) wouldn't be completed in time for production delivery to start in early 1943, a meeting was held by **Wa Pruef** on 28 July 1942 to discuss the rapid design of a **Zwischenloesung** (expedient design) **Selbstfahrlafette** that could be produced by early 1943. Alkett-Borsigwalde was awarded contracts to quickly design a **Zwischenloesung Fahrgestell** (chassis) using components available from **Pz.Kpfw.III** and **IV** chassis.

As revealed in Speer's conference with Hitler on 2 October 1942, Alkett quickly completed the design: *Hitler was pleased with the model of the Selbstfahrlafette with the same parts utilized for both the 8.8 cm Pak 43 and 15 cm s.F.H. on a Panzer IV Fahrgestell. He awaits compliance with the plans for delivery of 100 of each by 12 May 1943.*

As guided by **Wa Pruef**, Alkett designed a chassis with the same hull width as the **Pz.Kpfw.III** (wider than the **Pz.Kpfw.IV**). Therefore, drive train components selected to fit this hull width and shape were adopted from the **Pz.Kpfw.III**, including the **Kettenantrieb** (final drives and sprockets), **Lenkgetriebe** (steering unit with drive shafts), and the **SSG 77 Schaltgetriebe** (transmission) that was already paired with the steering unit. The rest of the drive train was adopted from the **Pz.Kpfw.IV**, including the **HL 120 TRM Motor**, **Luefter** (cooling fans), **Kuehler** (radiators), and **Schalldaempfer** (muffler) from the **Ausf.F**. The suspension components were also adopted from the **Pz.Kpfw.IV**, including the **Stuetzrolle** (return rollers), **Rollenpaare** (pairs of roadwheels), **Gleisketten** (tracks with 108 links), and **Leitrad** (idler for track tension adjustment). The engine was moved forward to the center of the hull and decked over with a reinforced platform for mounting the gun near the center of gravity for the vehicle. This also provided the crew with a more convenient, better sheltered space to serve the gun.

As described in the manual D653/42, this chassis has several unique features, including:

o **Kuehlluftfuehrung** (cooling air path): Cooling air entering the opening on the left side is pulled through the radiators and passes over the engine and blown out the opening on the right side by fans. Adjustable dampers on the left opening are operated from the driver's compartment. A damper on the right side to seal off the exhaust opening is operated by two hand knobs located below the **M.G.-Halter** in the fighting compartment. A damper to regulate the **Kampfraumheizung** (fighting compartment heater) is operated by a crank handle.

o A **Schwungkraftanlasser** (inertial starter) is mounted on the left side of the engine and is connected by a drive shaft to an **Andrehklaue** (starter claw) mounted on the rear firewall. The inertial starter is spun up by the crew using a hand crank in the fighting compartment.

o Fuel is stored in two tanks (total capacity of 600 liters) below the floor of the fighting compartment. The filler openings are accessible from the fighting compartment. A ventilation hole in the filler cap must always be kept open and clean. Otherwise, fuel supply to the engine will be disrupted. During preparation for combat, the drain port under the fuel tanks is to be opened to drain any fuel out of the **Panzerwanne** (armor hull). It is to be closed only when fording.

o A **Kuehlwasserheizgeraet Bauart Fuchs** (cooling water heater model Fuchs) is mounted on the left side with an access hole covered by a lid. The lid is pivoted out of the way to mount a **Loetlampe** (blow torch) for heating the coolant.

At 10 mm thick the armor protection on the superstructure and gun shield was designed to be proof only against shell fragments, but not proof against **7.92 mm SmK** (AP bullets) at close range. Originally, all the plates on the hull were to have been made out of **SM-Stahl** that was 20 mm thick on the sides and rear and 50 mm thick on the hull front. However, apparently to save weight, 30 mm face-hardened plate was actually used on the hull front.

Including the upper carriage, the **8.8 cm Panzerjaegerkanone 43/1 (L/71)** abbreviated **8.8 cm Pak 43/1 (L/71)** mounted in the **Panzerjaeger Hornisse (Fahrgestell Pz.Kpfw.III/IV)**, was identical to the **8.8 cm Pak 43/41** towed gun. The shape of the gun shield was changed to a curve that could maintain a closer fit with the superstructure sides as the gun was traversed. The **Rohrvorholer** (recuperator) was mounted above the gun tube, the **Rohrbremse** (recoil cylinder) below the gun, and

Above: When Hornisse were first completed by Alkett in February 1943 they had an adjustable cover over a slit in the gun shield for the direct-fire telescopic gunsight. These first Hornisse did not have a port in the hull side for a Kuehlwasserheizgeraet Bauart Fuchs (cooling water heater model Fuchs). (NA)

two **Ausgleicher** (counterbalance cylinders) were mounted on each side of the gun. The gun's elevation arc was limited from -5 to +20 degrees and traverse limited to 30 degrees (15 degrees to the right or left).

The **1.Serie 8.8 cm Pak 43/1 (L/71) Hornisse** had two gun sights: (1) the **Zieleinrichtung 43 SVo (Seitenvorhalt, optisch)** with **Zielfernrohr 3 x 8°** (telescopic gun sight with 3 x magnification and 8 degree field of view) for direct fire with **Panzergranaten** (range drums for **Pzgr.39** and **Pzgr.40** were mounted on the left side of the mount) and (2) the **Zieleinrichtung 34** with **Rundblickfernrohr 32** or **36** were for indirect fire with **Sprenggranaten** from covered firing positions. The center of the **Z.F.3x8** was 552 mm to the left and 82 mm below the center of the gun, while the center of the **Rbl.F.32** was 412 mm to the left and 265 mm above the center of the gun. A **Sehschlitz** (vision slit) for the **Zieleinrichtung 43 Svo** was cut out of the gun shield and covered with a hinged lid. The opening could be adjusted with a hand grip and secured by a **Sperrklinke** (safety catch).

Four different types of ammunition could be fired from the **8.8 cm Pak 43/1** using electrical detonation, the **8.8 cm Pzgr.39** weighing 10.2 kg with a muzzle velocity of 1000 m/s, the **8.8 cm Pzgr.40** weighing 7.3 kg with a muzzle velocity of 1140 m/s, the **8.8 cm Sprgr.** weighing 9.4 kg with a muzzle velocity of 700 m/s, and the **8.8 cm Gr.HL** weighing 7.62 kg with a muzzle velocity of 600 m/s. A total of only 16 rounds were stowed in ammunition bins (8 on each side), but it was intended that an additional 24 rounds be stowed in packing cases on the floor.

Official Names

The official names for this **Panzerjaeger** changed with time as follows:

o **s.Sfl. auf Pz.Kpfw.III/IV Fahrgestell, Hornisse mit 8.8 cm Pak 43**
(21Jan43 by **Hauptausschuss Panzerwagen u. Zugmaschinen**)

o **8.8 cm Panzerjaeger IVb (Hornisse)**
(26Jan43 by **Gen.St.d.H. Org.Abt.**)

o **Panzerjaeger "Hornisse" (Sd.Kfz.164) with 8.8 cm Panzerjaegerkanone 43/1 (L/71)**
(30Jan43 in **K.St.N.1148b**)

o **Panzerjaeger III/IV "Hornisse" fuer 8.8 cm Pak 43/1 (Sd.Kfz.164)**
(6Aug43 by **In 6**)

o **8.8 cm Panzerjaeger 43/1 (L/71) Hornisse**
(15Oct43 in **D 653/42**)

o **Panzerjaeger "Hornisse" fuer 8.8 cm Pak 43/1 (Sf) (Sd.Kfz.164)**
(1Nov43 in **K.St.N.1148b**)

o **Nashorn - Suggestivname fuer 8.8 cm Pak auf Fgst.III/IV**
(29Nov43 Hitler, 1Feb44 **OKW**, 27Feb44 **OKH**)

o **Pak 43 (Sf) (Hornisse)**
(1Mar44 in **D9020/46**)

o **Pz.Jaeg. "Nashorn" fuer 8.8 cm Pak 43/1 (Sd.Kfz.164)**
(6Jul44 by **In 6**)

o **8.8 cm Pak 43/1 Sf. "Nashorn" (frueher "Hornisse")**
(Sep44 by **Gen.Insp.d.Pz.Tr.**)

o **8.8 cm Pak 43/41 (L/71) auf Fgst. Pz.Kpfw.III u. IV (Sf.) (Nashorn) (Sd.Kfz.164)**
(Oct44 by **Waffenamt**)

Right: As shown on this second Hornisse (Fgst.Nr.310002 chalked on the front) completed by Alkett in February 1943, initially the opening for the brake vents were protected by an armor guard from a Pz.Kpfw.III Ausf.F. (KHM)

Production

As recorded in the **Adolf Hitler Panzerprogramm - Programm II/III** from the **Hauptauschuss Panzerwagen und Zugmaschinen** and signed by Rohland in Dusseldorf on 22 January 1943, assembly of the **s.Sfl. auf Pz.Kpfw.III/IV-Fahrgestell, Hornisse mit 8.8 cm Pak 43** was to occur at Alkett (Berlin) and Stahlindustrie (Duisburg). Alkett was to deliver the first 10 in January, 20 in February, 30 in March, and continue at the rate of 30 per month, for a total of 420 by the end of March 1944. Stahlindustrie was to start later with the first 5 completed by the end of May, 10 in June, 15 in July and continue at the rate of 15 per month, for a total of 150 by the end of March 1944.

On 6/7 February 1943, in Speer's conference with Hitler: *Even though Hitler basically views the **ungepanzerten Selbstfahrlafette fuer Pak** only as an expedient or interim design, the 100 **Pak 8.8 auf Fahrgestell III/IV (Hornisse)** are to be completed by 12 May 1943. Plans to continue production at the rate of 45 per month after May should be reduced to 20 per month to support **Hummel** production.*

Stahlindustrie reported in their delivery plan for May dated 3 May 1943: *Delivery of **Hornissen** is not to occur by us at Stahlindustrie because guns have been secured only for Alkett production. Therefore, Stahlindustrie will only complete **Hummel** for the next few months.*

As reported by **Wa J Rue (WuG)** on 11 May 1943: *The last 40 **Panzerjaeger Hornisse** with **8.8 cm Pak 43/1 (L/71)** are scheduled to be completed in September 1943. Production of the new **8.8 cm Pak 43** is to occur at the rate of 40 per month from September to January 1944 for mounting on **Panzerjaeger Hornisse** that are to be completed at the rate of 40 per month from October 1943 to February 1944.*

On 2 July 1943, **Wa J Rue (WuG)** reported: *A total of 500 **Hornissen** are to be delivered. It must still be clarified as to whether in future the new **8.8 cm Pak 43** will be mounted or if 173 **8.8 cm Pak 43/1** will have to be retained from next few month's production. Production of the **8.8 cm Pak 43/41** runs out at the end of 1943.* 163 **Pak 43/1** had been completed for **Hornisse** by the end of June, while another 200 (at the rate of 40 per month) had been scheduled to be completed by the end of November.

Plans to install a new **Pak 43** had been dropped by the end of July, and plans were made to escalate production of **Pak 43/41** to 60 per month so that all 500 would be completed for **Hornissen** by the end of December 1943. However, due to a bombing raid on Henschel, only 20 were completed in August 1943, and production of the **8.8 cm Pak 43/1** for the Hornisse stretched out to May 1944.

On 4 November 1943, the **Wa A/Wa J Rue (WuG)** reported to the **Generalinspekteur der Panzertruppen**: *284 **Hornissen** were completed by 31 October. The rest of the 216 **Hornissen** (500 in total) are scheduled to be completed at the rate of 40 per month from November through March 1944, with the last 16 completed in April 1944.*

On 21 November 1943, Generalleutnant Schneider reported: *The poor results achieved by **8.8 cm Pak 43/41** and **Hornisse** in action are due to sight misalignment caused by driving. In trials conducted by the **Waffenamt** there weren't any problems with the sights. The **Hornisse** should no longer be produced as an expedient solution.* Oberst Thomale (**Gen.Insp.d.Pz.Tr.**) requested: *Hornisse production should continue until it can be replaced by the **Panzerjaeger "Panther"**, which is to be delivered starting in February 1944. As an expedient, the **Hornisse** is better than having no **8.8 cm Pak Sfl.** in action.*

Then Alkett was hit by a heavy bombing raid in November 1943. This resulted in production being switched over to Deutsche Eisenwerke A.G. at their Stahlindustrie Werk in Duisburg and also their assembly plant in Teplitz-Schoenau. Already by 22 May 1943, Deutsche Eisenwerke had been awarded **Auftrag SS 210-8911/43** for assembly of 2064 **Hummel, Hornisse,** and **Muni**

As recorded in the weekly reports from the D.E.W. Werk Teplitz-Schonau:
28Feb44 Report No.15 - *25 guns and upper carriages have arrived for **Hornissen**.*
18Mar44 Report No.17 - *12 superstructure supports from Alkett were completed here.*
1Apr44 Report No.18 - *Preparations for **Hornisse** assembly will begin on 3 April. 10 **Hornissen** chassis arrived from Duisburg. Radio sets haven't been installed in eight chassis. The last railcar with the rest of the parts for the first 20 **Hornisse** superstructure should be on the way here from Alkett.*
3May44 Report No.20 - *5 **Hornisse** were shipped. 20*

Production Statistics			
	1943	1944	1945
Jan	0	0	12
Feb	14	25	3
Mar	30	0	1
Apr	41	20	
May	35	24	
Jun	35	6	
Jul	44	3	
Aug	16	31	
Sep	27	12	
Oct	42	7	
Nov	24	5	
Dec	37	0	
Total	345	133	16

Hornisse were presented for inspection, of which 14 were accepted. 25 to 30 *Hornissen* are to be completed in May. 9May44 Report No.21 - *Instead of 14, 20 should have been reported as accepted in weekly report No.20. 18 of these Hornissen were shipped by today. Up to now, 11 chassis have arrived for May production.*

The **vorlaeufiges Richtwertprogramm IV** (schedule for production in the 4th fiscal year) dated 4 May 1944 reveals that Alkett had dropped completely out of **Nashorn** production. In this document, Stahlindustrie is scheduled to complete 30 in April, 30 in May, and the last 40 in June 1944 for a total of 100. This schedule was changed in **vorlaeufiges Richtwertprogramm IV** dated 14Jul44 to 12 **Nashorn** in April, 24 in May, 5 in June, 30 in July, 30 in August, and 29 in September for a total of 130 from Stahlindustrie.

Further disruptions occurred causing **Nashorn** production to drag out to the end of the war. In the **Panzermontageprogramm** (armored vehicle assembly program) dated 30 January, Stahlindustrie was scheduled to complete 9 in January, and the last 2 in February 1945.

On 14 March 1945, a meeting was held by the **Generalinspekteur der Panzertruppen** to settle development questions including production of an **8,8 cm-Pak Waffentraeger auf Fahrgestell "Hummel"**: *430 "Hummel" were included in the Auslaufprogramm dated 13Feb45, with priority given to 250 "Hummel" Sf. mit 15 cm s.F.H. for the Artillerie. Therefore, 180 "Hummel" chassis are available on which the 8.8 cm Pak als Sf. could be mounted and therefore the same number of 180 Zugkraftwagen could be saved for towing vehicles. Because the artillery requirements for 250 "Hummel" have priority, and in the best case only sufficient material can be recovered out of Duisburg to complete 199 to 200 vehicles, there is no chance that the Gen.Insp.d.Pz.Tr. will get any of the remaining chassis for 180 Pak 8.8 cm Sf.*

Modifications

The following modifications were introduced on new **Panzerjaeger "Hornisse"** in the assembly plants during the production run:

o A curved armor shield mounted on the recuperator cylinder was added in February 1943

o The armor casting for the brake vents was changed from the **Pz.Kpfw.III Ausf.F** design to a larger armor casting with a larger opening in March 1943.

o Brackets on the hull rear to stow the gun cleaning rods were welded on closer together and aiming stakes were no longer carried starting in March 1943.

o The brackets welded to the glacis plate for stowing tow cables were relocated and a U-shaped bracket added to hold the ends, starting in March 1943.

o A hole covered by a armor cap was cut into the left hull side for a **Fuchs-Geraet** to preheat engine coolant with a blow torch for cold weather starts, starting in March 1943.

Above: Four Hornisse completed by Alkett in February/March 1943 that were issued to the first unit to be outfitted with these tank destroyers, schwere Panzer-Jaeger Abteilung 560. They all have initial production features - Pz.Kpfw.III Ausf.F brake vents and Pz.Kpfw.III Ausf.H sprockets. (KHM)

o Starting in March 1943, the sprocket wheel originally designed for the **Z.W.38 (Pz.Kpfw.III Ausf.E)** was intermittently mounted on **Hornisse** instead of the cast sprocket wheel with 6 holes designed for the **Z.W.39 (Pz.Kpfw.III Ausf.H)**

o The first 50 guns had two sights (a **Zieleinrichtung 43 SVo** with **Zielfernrohr 3 x 8°** for direct fire and a **Zieleinrichtung 34** with **Rundblickfernrohr 32** or **36** for indirect fire. In April 1943, starting with the 51st **8.8 cm Pak 43/1** the sight was changed to a **Zieleinrichtung 37** with a **Sfl. Z.F.1a** periscope for both direct and indirect fire. The slot cut into the left gun shield was no longer needed. Many additional gun shields had already had the slot cut before this modification went into effect. These were welded closed on gun shields still being mounted as late as August 1943.

o Starting in April 1943, the "box" mounted below the panniers on both sides at the rear was dropped. Tow hooks were then welded to both sides at the rear and the internal lifting eye (previously bolted on the inside to the hull rear) was moved forward and welded to the hull side.

o Starting in April 1943, the **Abgasschalldaempfer** (exhaust muffler) was dropped and brackets welded to the hull rear to stow two spare roadwheels.

o The second **Tarnscheinwerfer** (headlight with blackout cap) mounted on the right track guard was dropped starting in May 1943. The armor guard for the electrical penetration was still welded to the glacis plate on the right side at least as late as August 1943.

o Starting in May 1943, the forward **Rohrstuetze** (clam shell travel lock), which had to be unbolted by a crew member being exposed to enemy fire, was replaced by a new design which could be released from the inside the fighting compartment. The **Gefechtszurrung** (combat travel lock) at the rear of the gun, mounted inside the fighting compartment, was also dispensed with in May 1943.

o A "loop" over the recuperator cylinder (bolted to the two halves of the gun shield) was added as a strengthener, starting in May 1943.

o The brackets for stowing the tow cables on the glacis plate were relocated farther to the right with the U-bracket being welded lower than the Z-brackets, starting by August 1943.

o Several different types of track links were mounted on the **Hornisse**. The **Winterketten** (tracks with side extensions) were introduced in the Winter of 1942/43, an **Einheitsketten** (standardized track link) with flat face were introduced in the Summer of 1943, **Gleitschutzpickeln** (chevrons) were cast onto the track face starting in the Fall of 1943, and wider **Ostketten** were introduced for use on the Eastern Front in June 1944.

o In the Fall of 1943, stamped hub caps for the roadwheels were introduced reflecting an internal change to the wheel bearings.

o At some point in 1944, a 230-mm-diameter cap (similar to the **Hummel** but with five retaining bolts) for the "**Fuchs-Geraet**" hole on the left hull side replaced the cast cap retained by two caps.

All the hulls for the **Panzerjaeger "Hornisse"** had been completed by Witkowitzer Bergbau- und Eisenhuetten by the end of 1943. This may account for some modifications not being introduced on the **Hornisse/Nashorn** that were introduced in the parallel **Hummel** production run (such as, reduction from 4 to 3 return rollers per side and driver's compartment extended across the full hull width).

Left:
Initially, the Hornisse had a rear travel lock for the 8.8 cm Pak 43/1 (visible through the open door with its large hand wheel) and an engine exhaust muffler mounted across the rear. The antenna with an armor pot mounted in the upper right superstructure rear, reveals that this is a Befehls-Hornisse with the longer range Fu 8 radio set. (KHM)

Organization

A **K.St.N.** (table of organization) had yet to be created for a **Hornisse Kompanie**, when the first unit was ordered to be formed on 26 January 1943. Instead, the **4.Kp./Pz.Jg.Abt.41** was ordered to be formed with ten **8.8 cm Panzerjaeger IVb (Hornisse)** in accordance with **K.St.N.1148a Ausf.A** dated 1Dec42. The associated table of equipment (**K.A.N.1148a** dated 1Dec42) authorized one **Pz.Sfl.** in each **Kompanie** to be outfitted with an **Fu 8** radio set and all 10 **Pz.Sfl.** in the **Kompanie** to be outfitted with a **Funksprechgeraet "f"** radio set.

The first **K.St.N.** created specifically for "Hornisse" units was **K.St.N.1148b Behelf** dated 30Jan43 for a **Panzerjaegerkompanie "Hornisse" (8.8 cm Pak 43)**, which was available when the next two companies were ordered to be formed on 6 and 10 February 1943. Each **Kompanie** was still authorized to have a total of only 10, one **Panzerjaeger "Hornisse" (Sd.Kfz.164)** was assigned to the **Gruppe Fuehrer** (company headquarters), two **Panzerjaeger "Hornisse" (Sd.Kfz.164)** to the **1., 2.,** and **3.Zug** (1st, 2nd, and 3rd Platoon, and three **Panzerjaeger "Hornisse" (Sd.Kfz.164)** were retained as a **Geraetreserve** (equipment reserve).

It wasn't until 25 March 1943 that orders were cut to consolidate the first three companies into a single battalion (**schwere Panzerjaeger-Abteilung 560 (Hornisse)**) to be formed with a **Stab** (**K.St.N.1106b** dated 30Mar43) and **Stabs-Kompanie** (**K.St.N.1155b** dated 30Mar43). On 1 May 1943 orders were cut authorizing the three companies in **s.Pz.Jg.Abt.560** to reorganize in accordance with **K.St.N.1148b** dated 1Apr43 with 14 **Hornissen** in each company. This **K.St.N.1148b Panzerjaegerkompanie "Hornisse" (8.8 cm Pak 43)** dated 1Apr43 was the **Ersatz** (replacement) for **K.St.N.1148b Behelf** dated 30Jan43.

The second battalion (**schwere Panzerjaeger-Abteilung 655**) was ordered to be formed on 14 April 1943 with three companies (**Kompanie 521, Kompanie 611,** and **Kompanie 670**). These three companies were ordered to be organized in accordance with **K.St.N.1148b** dated 1Apr43 authorizing each company to have 14 **Hornisse**.

On 25 April 1943, the third battalion (**schwere Panzerjaeger-Abteilung 525 (8.8 cm Pz.Jaeg.IVb Hornisse)** was ordered to be formed with an **Abteilung Stab** organized in accordance with **K.St.N.1106b** dated 30Mar43, a **Stabs-Kompanie** organized in accordance with **K.St.N.1155b** dated 30Mar43, and three **Kompanien** organized in accordance with **K.St.N.1148b** dated 1Apr43.

K.St.N.1155b published on 1 May 1943 for a **Stabskp. einer Panzerjaeger-Abteilung zu "Hornisse"** authorizing the headquarters company to possess three **Panzerjaeger "Hornisse" (Sd.Kfz.164)** was the **Ersatz** (replacement) for the **K.St.N.1155b Behelf** dated 30Mar43.

In August and September 1943 the fourth and fifth battalions (**schwere Panzerjaeger Abteilung 93 and 519 (Hornisse)**) were ordered to be formed with an **Abteilung Stab** organized in accordance with **K.St.N.1106b** dated 30Mar43, a **Stabs-Kompanie** organized in accordance with **K.St.N.1155b** dated 28Aug43, and three **Kompanien** organized in accordance with **K.St.N.1148b** dated 28Aug43. **K.St.N.1155b** dated 28Aug43 for a **Stabskp. einer Panzerjaegerabteilung zu "Hornisse"** authorized this headquarters company to have three **Panzerjaeger "Hornisse" (Sd.Kfz.164)**. **K.St.N.1148b** dated 28Aug43 for **Panzerjaegerkompanie "Hornisse" (8.8 cm Pak 43)** authorized two **Panzerjaeger "Hornisse" (Sd.Kfz.164)** for the **Gruppe Fuehrer**, and four **Panzerjaeger "Hornisse" (Sd.Kfz.164)** each for the **1., 2., und 3.Zug**.

As revised on 1 November 1943, **K.St.N.1155 b** for a **Stabs Kp. s.Pz.Jaeg.Abt. "Hornisse"** still authorized the headquarters company to posses three **Panzerjaeger "Hornisse" fuer 8.8 cm Pak 43/1 (Sf) (Sd.Kfz.164)** and **K.St.N. 1148b** for a **Panzerjaegerkompanie "Hornisse" (8.8 cm Pak 43)** still authorized each company to possess 14 **Panzerjaeger "Hornisse" fuer 8.8 cm Pak 43/1 (Sf) (Sd.Kfz.164)**. The last complete battalion (**schwere Panzerjaeger-Abteilung 88**) was ordered to be formed on 3 December 1943 with an **Abteilung Stab** organized in accordance with **K.St.N.1106b** dated 30Mar43, a **Stabs-Kompanie** organized in accordance with **K.St.N.1155b** dated 1Nov43, and three **Kompanien** organized in accordance with **K.St.N.1148b** dated 1Nov43.

As announced in the **Nachrichtenblatt der Panzertruppen** dated April 1944: **K.A.N.1148b** dated 1Nov43 authorized one **8.8 Pak - Sf. "Hornisse"** to be outfitted with both a **Fu 8** and a **Fu 5** radio set and the other 13 **8.8 cm Pak - Sf. "Hornisse"** in the company to be outfitted with one **Fu 5** radio set. **K.A.N.1155b** for the **Stabskp.** dated 1Nov43 authorized two **8.8 Pak - Sf. "Hornisse"** to be outfitted with both a **Fu 8** and a **Fu 5** radio set and one **8.8 cm Pak - Sf. "Hornisse"** to be outfitted with only one **Fu 5** radio set.

Schwere Panzerjaeger-Abteilung (Hornissen) 560

Originally the plan was to issue 10 **Hornissen** to a single **Kompanie** in the **Panzerjaeger-Abteilung** in a **Panzer-Division** to provide them with a core of self-propelled anti-tank guns capable of knocking out any enemy tank that could be encountered at long range. Orders from the **Gen.St.d.H. Org.Abt.** were issued to form the following units on:

<u>26Jan43</u> to form the **4.Kp./Pz.Jg.Abt.41** with 10 - **8.8 cm Panzerjaeger IVb (Hornisse)** in accordance with **K.St.N.1148a Ausf.A** dated 1Dec42 to be combat ready 15Feb43.

<u>6Feb43</u> to form the **4.Kp./Pz.Jg.Abt.42** with 10 - **8.8 cm Panzerjaeger IVb (Hornissen)** in accordance with **K.St.N.1148b** dated 30Jan43 to be combat ready 28Feb43

<u>10Feb43</u> to form the **4.Kp./Pz.Jg.Abt.61** with 10 - **8.8 cm**

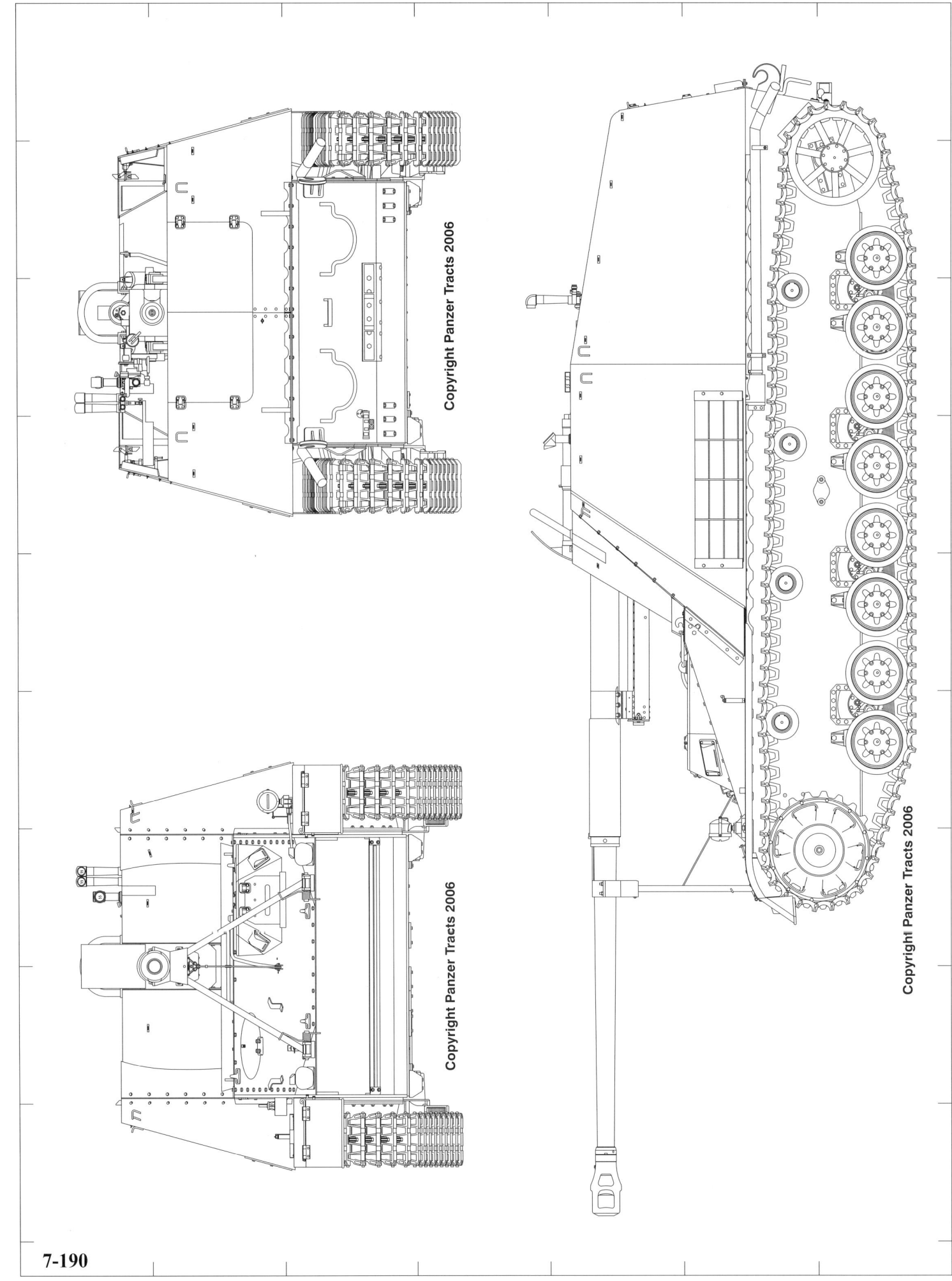

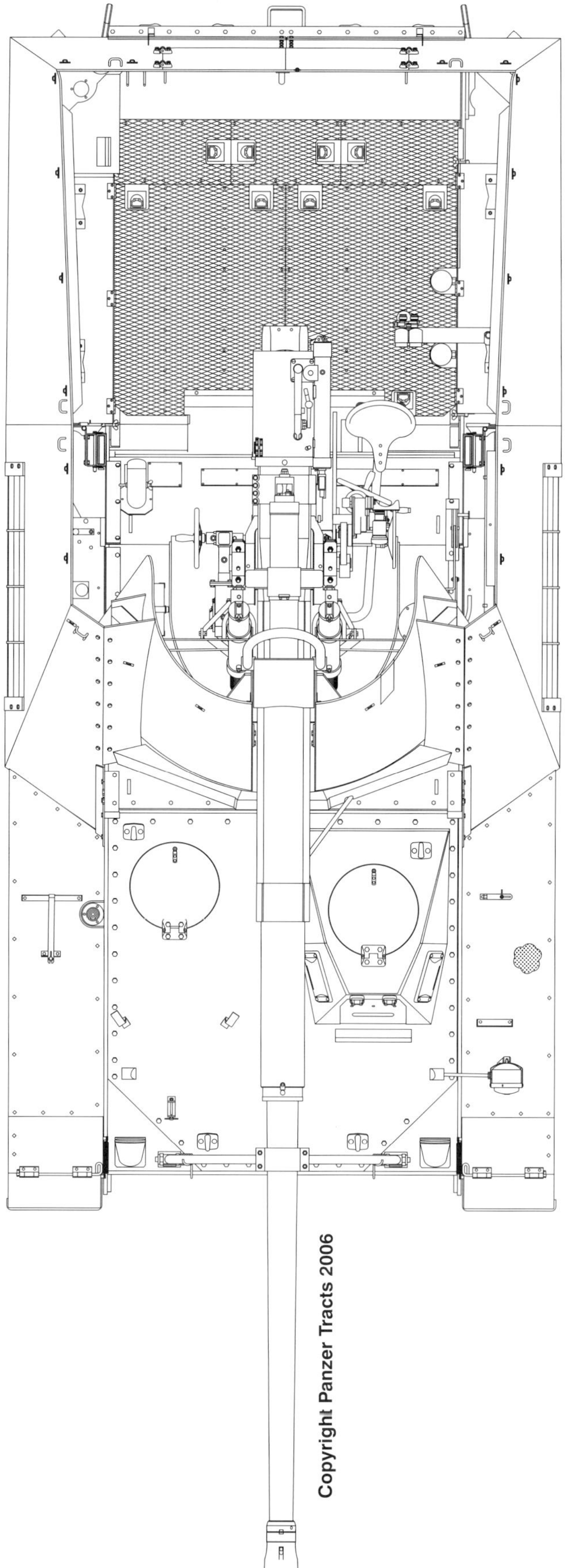

A Panzerjaeger "Hornisse" as completed by Alkett in June 1943.

This and Opposite Page: This Hornisse, issued to a Pz.Jg.Ers.u.Ausb.Abt. (training unit), has the features of one completed by Alkett in June/July 1943, including the reinforcing loop for the gun shield, slit in the gun shield for a direct-fire sight welded shut, remotely lowerable front gun travel lock, single headlight, larger cast brake vents, ends of exhaust pipes bent to reduce dust, spare roadwheels mounted on the hull rear, and a central step on the hull rear.
(TTM)

Above: An armor guard covered the opening in the hull side for the Kuehlwasserheizgeraet Bauart Fuchs (engine coolant heater model Fuchs), used for cold weather starts. (TTM)

Right:
The Gefechtszurrung (combat travel lock) at the rear of the gun, mounted inside the fighting compartment, was dispensed with in May 1943 after the remotely lowerable front gun travel lock was introduced.
 (TTM)

Panzerjaeger IVb (Hornissen) in accordance with **K.St.N.1148b** dated 30Jan43 to be combat ready 15Mar43

On 15 February 1943, **Gen.St.d.H.** reported: Instead of as previously planned, the first three **Kompanien 8.8 cm Pak 43 (Sfl.)** will be sent and incorporated into the **1., 6., and 7.Panzer-Divisionen**. The previously created **Hornissen** units will be renamed as follows:
4.Kp./Pz.Jg.Abt.41 into **3.Kp./Pz.Jg.Abt.37**
4.Kp./Pz.Jg.Abt.42 into **4.Kp./Pz.Jg.Abt.41**
4.Kp./Pz.Jg.Abt.61 into **4.Kp./Pz.Jg.Abt.42**

At the end of February 1943, the **1.Panzer-Division** reported that the **3.Kp./Pz.Jg.Abt.37 (Sfl. Hornissen)** was being formed currently in Wuensdorf and wasn't expected to join the division until 28 March. On 20 March 1943, **Gen.St.d.H.** ordered that the three **s.Pz.Jg.Kp. (Hornissen)** assigned to the **1.Panzer-Division** be concentrated into an **Abteilung** and the **Kompanie** be filled to 15 **8.8 cm Pz.Jg.IVb (Hornisse)** each.

On 25 March 1943, **Gen.St.d.H. Org.Abt.** ordered **schwere Panzerjaeger-Abteilung 560 (Hornisse)** to be formed with a **Stab** (**K.St.N.1106b** dated 30Mar43) and **Stabs-Kompanie** (**K.St.N.1155b** dated 30Mar43). The three previously created "Hornisse" companies were to be renamed as follows: **3.Kp./Pz.Jg.Abt.37 (1.Pz.Div.)** as **1.Kp./Pz.Jg.Abt.560**, **4.Kp./Pz.Jg.Abt.41 (6.Pz.Div.)** as **2.Kp./Pz.Jg.Abt.560**, and **4.Kp./Pz.Jg.Abt.42 (7.Pz.Div.)** as **3.Kp./Pz.Jg.Abt.560**. The **Abteilung** was to be attached to the **1.Panzer-Division**. After being organized, the **Stab** was to be sent to the **1.Panzer-Division** in **O.B.West**. After 29 March, the three **Kompanien** were to be transferred to **O.B. West**.

In a status report dated 28 April 1943 on the **s.Pz.Jg.Abt.560 (Hornisse)**, it was reported that the **Kompanien** had been created in accordance with **K.St.N.1148b** dated 1Jan43 and therefore had 10 **Hornissen** in each **Kompanie**, while the current **K.St.N.1148b** dated 1Apr43 calls for 14 **Hornissen** in each **Kompanie**. On 1 May 1943, orders were issued to reorganize the three **Kompanien** in accordance with **K.St.N.1148b** dated 1Apr43.

s.Pz.Jg.Abt.560 Operational Status Reports				
Date	Avail	Oper	Repair	Location
31Jun43	45	37	8	H.Gr.Sued
31Jul43	45	39	6	H.Gr.Sued
31Aug43	31	18	13	H.Gr.Sued
30Sep43	35	10	25	H.Gr.Sued
31Oct43	39	8	31	H.Gr.Sued
30Nov43	44	13	31	H.Gr.Sued
31Dec43	31	14	17	H.Gr.Sued
31Jan44	26	12	14	
1Mar44	14	4	10	LVII Pz.K.

The first six **Hornissen** had been issued in February, followed by 24 in March. It was only in May 1943 that an additional 15 **Hornissen** were actually sent from the **Heereszeugamt** (6 on 14 May and 9 on 17 May 1943) to fill the unit to its total authorized complement of 45.

As ordered on 21 April 1943, after 22 April, **s.Pz.Jg.Abt.560 (Hornisse)** was to be transferred to Charkow and attached to **H.Gr.Sued**. Most of the **Abteilung** was to arrive on 2 May. On 2 July 1943, **s.Pz.Jg.Abt.560** was assigned to the **XXXXII A.K.** under **Armee Abteilung Kempf** to assist in protecting the flank during the **III. Pz.K.** attack during Operation **"Zitadelle"**. As ordered by Generalfeldmarschall von Manstein, the **s.Pz.Jg.Abt.560** was employed to secure the roads and took up position on 6 July 1943.

They didn't lose a single **Hornisse** in July that couldn't be repaired. Still attached to the **XXXXII A.K.** with its **Kompanien** assigned to support the **39., 161., and 282.Infanterie Divisionen**, **s.Pz.Jg.Abt.560** reported 14 total losses in August. To replace their losses 5 Hornisse were sent as replacements from the **Heereszeugamt** on 30 September, 5 on 31 October, 5 on 28 November, and the last 4 on 3 February 1944. At the end of 1943, **s.Pz.Jg.Abt.560** reported that they had knocked out a total of 251 enemy tanks.

On 4 February 1944, **s.Pz.Jg.Abt. (Hornisse) 560** was ordered to be pulled out as soon as possible from **H.Gr.A** and transferred to Mielau for reorganization with **Jagdpanther**. On 1 March 1944, they reported losing a total of 16 **Hornisse** while fighting under the **LVII Pz.K.** By the end of April 1944, the 560th was in Mielau reorganizing as a **Jagdpanther** unit.

Schwere Panzerjaeger-Abteilung 655 (previously named "Stalingrad")

On 20 March 1943, the **Gen.St.d.H.** decided to form **s.Pz.Jg.Abt. "Stalingrad"** with a **Stab** and three **Kompanien** each with 15 **8.8 cm Pz.Jg.IVb (Hornisse)**. As ordered on 14 April 1943, the **Stab Pz.Jg.Abt. "Stalingrad"** was renamed **schwere Panzerjaeger-Abteilung 655**. Formed from the surviving remnants of **Pz.Jg.Abt.521, 611, and 670**, instead of being numbered 1., 2., and 3., the three **Kompanien** retained the designations from the parent units as **Kompanie 521**, **Kompanie 611**, and **Kompanie 670**. The **Stab** was to be organized in accordance with **K.St.N.1106b** dated 30Mar43, and each of the three **Kompanien** in accordance with **K.St.N.1148b** dated 1Apr43.

In a status report dated 5 May, **s.Pz.Jg.Abt.655** with three **Kompanien (521, 611, 670)** with 14 **8.8 cm Hornissen** each was to be combat ready by 15 May 1943. The unit had been issued 35 **Hornisse** in April and sent the last 10 from the **Heereszeugamt** in May (six being sent on 7 May and the last four on 14 May) to complete its total authorized establishment of 45.

On 25 April 1943, even though the **Abteilung** was not combat ready, it was ordered to be sent from Spremberg to **H.Gr.Mitte** starting on 30 April. On 6 May, it was reported that the **Abteilung** was expected to arrive in Mogilew starting on 8 May. Any missing equipment was to be sent after. **Auffrischungsstab Mitte** reported the arrival of the **s.Pz.Jg.Kp.521** in Minsk on 12Aug43.

To replace losses, **s.Pz.Jg.Abt.655** was sent the following replacement **Hornissen** from the **Heereszeugamt**: 8 on 27 July, 5 on 31 October, 5 on 3 November, 5 on 1 December, and 10 on 10 March 1944.

As ordered on 15April 1944, **Pz.Jg.Kp.521** was renamed **1.Kp./Pz.Jg.Abt.655**, **Pz.Jg.Kp.611** renamed **2.Kp./Pz.Jg.Abt.655**, **Pz.Jg.Abt.670** was renamed **3.Kp./Pz.Jg.Abt.655**.

Leaving behind an **Einsatz-Kompanie** with the remaining **s.Pz.Jg.**, on 19 August 1944, **s.Pz.Jg.Abt.655** was ordered to be pulled out of **H.Gr. Nordukraine** and be sent to Mielau for conversion to a **Jagdpanther** unit.

s.Pz.Jg.Abt.655 Operational Status Reports				
Date	Avail	Oper	Repair	Location
30Jun43	40	13	27	H.Gr.Mitte
31Jul43	36	24	12	H.Gr.Mitte
31Aug43	40	26	14	H.Gr.Mitte
30Sep43	36	20	16	H.Gr.Mitte
31Oct43	36	28	8	H.Gr.Mitte
30Nov43	45	20	25	H.Gr.Mitte
1Jan44		22	12	H.Gr.Mitte
31Jan44	47	9	38	H.Gr.Mitte
1Mar44	41	0	41	H.Gr.Mitte
1May44		45	9	253.Inf.Div.
1Jun44	48	47	1	253.Inf.Div.
1Jul44	48	45	3	26.Inf.Div.
1Aug44	25	14	7	

Above: Hornisse (Fgst.Nr.310042 completed by Alkett in late March/early April 1943) issued to s.Pz.Jg.Abt.655, has been backfitted with an remotely lowerable travel lock with a unique (field modification) quick release and extra armor added to the gun shield as an unauthorized unit improvisation. (KHM)

schwere Panzerjaeger-Abteilung 525

As ordered on 25 April 1943, **Panzerjaeger-Abteilung 525** was to be reorganized as a **schwere Panzer-Jaeger-Abteilung (8.8 cm Pz.Jaeg.IVb Hornisse)** with:

o **Abteilung Stab** organized in accordance with **K.St.N.1106b** dated 30Mar43

o **Stabs-Kompanie** organized in accordance with **K.St.N.1155b** dated 30Mar43

o three **Kompanien** organized in accordance with **K.St.N.1148b** dated 1Apr43

After filling the authorized personnel slots, the unit was to be transferred to the **O.B. West** sector, where it was to be outfitted with equipment and trained.

As reported on 5 May 1943, **s.Pz.Jg.Abt.525** organized with three **Kompanien** each with 14 **8.8 cm Hornissen** and a **Stabs-Kp.** with 3 **Hornissen** was to be combat ready by 15 June 1943.

On 15 May 1943, the **Abteilung** was ordered to be transferred from **O.B.West** to Coelquidan after 5 June 1943. **Kommandos** were to be sent to **H.Za.Magdeburg** to take over **Panzerjaeger Hornisse**. A total of 45 **Hornissen** were issued, with the first 3 sent from the **Heereszeugamt** on 22 May, followed by 7 on 2 June, 10 on 19 June, 10 on 1 July, 8 on 2 July, and the last 7 on 10 July 1943.

On 10 July 1943, **s.Pz.Jg.Abt. (Hornisse) 525** was ordered to be shipped out directly following completion of the movement of the **26.Panzer-Division** after 20 July 1943. On 15 July 1943, **Pz.Jg.Abt.525** was ordered to be attached to **Ob Sued** as **Heerestruppe** (independent unit) and was attached temporarily to the **26.Pz.Div.** on 17 July. Based on the current training status on 19 July, **s.Pz.Jg.Abt.525** was to be combat ready on 1 August 1943.

Initially sent to northern Italy, it was then sent south to oppose the Allied landings. In the monthly status report on 1 February 1944, **s.Pz.Jg.Abt.525** reported: Starting on 5 January, the **Abteilung** with **3.Kompanie**, and the rest of the **Abteilung** after 18 January 1944, has been continuously on the march and in combat on the Suedfront near Cassino and in the Nettuno-Ring. Four **Hornisse** were destroyed by artillery direct hits, and three more, damaged by artillery hits, will take over 14 days to repair.

On 1 March 1944, the **s.Pz.Jg.Abt.525** reported: Favorable effective fire can be achieved in suitable terrain. For example, a Sherman was knocked out at 2800 meter range. The crews have faith in their weapon. Thicker armor is desired. The first five replacements for losses left the **Heereszeugamt** on 26 May 1944.

On 28 August 1944 it was reported: The **s.Pz.Jg.Abt.(Hornisse)** in **O.B.Suedwest** had only 16 **Hornissen** (of which 3 were being repaired) on 1 August 1944. **Gen. Insp.d.Pz.Tr.** intends to convert this unit into a **gemischte. Jagdpanther-Abteilung** and requested that the **Abteilung** be transferred to Mielau. On 31 August 1944, the **1.Kompanie** from **s.Pz.Jg.Abt.525 (Nashorn)** was ordered to be sent to Mielau for rearming. After being equipped with **Jagdpanther**, the **Kompanie** was to return to **H.Gr.C**.

S.Pz.Jg.Abt.525 was sent another 20 replacements (15 leaving the **Heereszeugamt** on 13 September and 5 on 23 September) and recorded in their monthly status report dated 1 October 1944: With the 20 **Nashorn** that were newly issued, both of the **Kampf-Kompanien** in Italy are almost 100% operational. Of the 20, 15 **Nashorn** are already in Italy, with 6 are operational and 7 in repair. 5 **Nashorn** still haven't arrived in Italy.

An additional 20 **Nashorn** were sent as replacements to **s.Pz.Jg.Abt.525** (10 leaving the **Heereszeugamt** on 15 October and the last 10 on 9 November 1944). In their monthly status report dated 1 November 1944, **s.Pz.Jg.Abt.525** recorded: After issue of another 10 **Nashorn**, the **Abteilung** now has an excess of 11 **Nashorn**. As ordered by **A.O.K.10**, these are to be set aside as a **Geraete-Reserve** (equipment reserve).

\multicolumn{5}{c}{**s.Pz.Jg.Abt.525**}
\multicolumn{5}{c}{Operational Status Reports}
Date	Avail	Oper	Repair	Location
30Jun43	20	20	0	West
31Aug43	45	41	4	West
30Sep43	45	42	3	H.Gr.B
31Oct43	45	45	0	H.Gr.B
30Nov43	45	45	0	H.Gr.C
31Dec43	45	41	4	H.Gr.C
1Feb44	41	23	18	3.Pz.Gr.Div.
1Mar44	36	29	7	I.Fallsch.K.
1Apr44	36	32	4	I.Fallsch.K.
1May44	35	30	5	I.Fallsch.K.
1Jun44	23	3	20	26.Pz.Div.
1Jul44	16	3	13	A.O.K.14
1Aug44	16	13	3	162.Inf.Div.
1Sep44	4	4	0	162.Inf.Div.
1Oct44	15	8	7	1.Fsh.Jg.Div.
1Nov44	32	13	19	114.Jg.Div.
1Dec44	42	31	11	A.O.K.10
1Jan45	37	33	4	LXXIII A.K.
1Feb45	34	30	4	
1Mar45	34	32	2	A.O.K.10
1Apr45	34	32	2	LXXVI Pz.K.

Above:
This Hornisse (completed by Alkett in May 1943) was issued to s.Pz.Jg.Abt.525. It has only a single headlight but still has the initial design for the front travel lock, which needed to be released by a crew member climbing out onto the front in combat situations.
(KHM)

Right:
Another Hornisse with the 2.Kp./s.Pz.Jg.Abt.525 in Italy has the triangular base for the protected antenna mount in the right rear corner. A feature needed to mount the extra long range radio equipment to convert a Hornisse to a command vehicle for use by company or battalion commanders. (DT)

This and Opposite Page: The crew in this Hornisse (Fgst. Nr.310107 completed by Alkett in May 1943) are preparing for action. After removing the periscopic Sfl.Z.F.1a gunsight from the storage box and locking it in position (above), the gunner (left) is adjusting the dial on the Z.E.37 for the estimated range. The loader (above right) is removing a round from the stowage bin and loading it in the 8.8 cm Pak 43/1. Starting in May 1943, a loop was added to reinforce the two halves of the gun shield. The rear travel lock had been discontinued by the time this Hornisse was completed, but the mounting pads on the hull sides were still present. (BA 311/907/33 + 38 and BA 311/908/2 + 4)

7-199

schwere Panzerjaeger-Abteilung 93 (Hornisse)

On 1 August 1943, **Pz.Jg.Abt.93** with their **1., 2.,** and **3.Kp.** were ordered to be renamed **s.Pz.Jg.Abt.93 (Hornisse)** and reorganized as:

o **Abteilung Stab** organized in accordance with **K.St.N.1106b** dated 30Mar43
o **Stabs-Kompanie** organized in accordance with **K.St.N.1155b** dated 28Aug43
o three **Kompanien** organized in accordance with **K.St.N.1148b** dated 28Aug43

They were issued 45 **Hornissen**, with 16 being sent from the **Heereszeugamt** on 29 July, 10 on 4 August, 10 on 18 August, and the last 9 on 10 September 1943. Sent to **Heeres Gruppe A** on the Eastern Front in September 1943, the **s.Pz.Jg.Abt.88** were sent 5 replacement **Hornisse** from the **Heereszeugamt** on 30 November 1943, 10 on 30 January 1944, 5 on 3 May, 10 on 5 July, and the last 10 replacements on 1 August 1944.

S.Pz.Jg.Abt.519, ordered to return to their home base to be converted to **Jagdpanther** and **Sturmgeschuetz** hadn't arrived by 7 October, but was reported as being in Potsdam on 2 November.

s.Pz.Jg.Abt.93 Operational Status Reports				
Date	Avail	Oper	Repair	Location
31Aug43	36	34	2	West
30Sep43	45	41	5	H.Gr.A
31Oct43	34	11	23	H.Gr.A
30Nov43	34	20	14	H.Gr.A
31Dec43	36	28	8	H.Gr.A
31Jan44	36	34	2	
1Mar44		23	6	A.O.K.6
1Apr44		14	3	A.O.K.6
1Jul44	15	14	1	H.Gr.NU
1Aug44		22	3	

schwere Panzerjaeger-Abteilung 519 (Hornisse)

On 17 September 1943, the **s.Pz.Jg.Abt.519 (Hor-nisse)** was ordered to be transferred to the **Truppen Uebungs Platz** (troop training area) **Oldebrock (W.B.Hdl.)** to continue their reorganization under the **VIII.Res.Pz.Korps** as follows: an **Abteilung Stab** organized in accordance with **K.St.N.1106b** dated 30Mar43, a **Stabs-Kompanie** organized in accordance with **K.St.N.1155b** dated 28Aug43, and three **Kompanien** organized in accordance with **K.St.N.1148b** dated 28Aug43.

The first 16 **Hornissen** were sent from the **Heereszeugamt** on 31 October, followed by 23 on 6 November, and the last 6 on 18 November 1943 (for a total of 45).

Schwere Panzerjaeger-Abteilung 519 was ordered to be sent to **H.Gr.Mitte** on 7 December 1943. To replace losses, 5 **Hornissen** were sent as replacements from the **Heereszeugamt** on 22 March, followed by 5 on 1 April, and the last 5 on 12 June 1944.

S.Pz.Jg.Abt.519 remained with **H.Gr.Mitte** until **Gen.Insp.d.Pz.Tr.** requested that they be pulled out on 5 August 1944. With the exception of an **"Einsatz-Kompanie"** to remain with **H.Gr.Mitte**, **s.Pz.Jg.Abt.519 (Nashorn)** was ordered to be sent back to Mielau on 8 August 1944. On 10 August 1944, this **Einsatz-Kompanie** from the **s.Pz.Jg.Abt.519 (Nashorn)** was ordered to give away their **Panzerjaeger "Nashorn"** and return to their **Abteilung** in Mielau. An **Ausbildungs-Kommando** (training unit) was to be left behind with **Pz.A.O.K.3**. On 22 August 1944, the 519th was ordered to convert to **s.Pz.Jg.Abt.(Panther) 519** with **Jagdpanther**.

s.Pz.Jg.Abt.519 Operational Status Reports				
Date	Avail	Oper	Repair	Location
25Nov43	45	45	0	West
30Nov43	45	40	5	H.Gr.Mitte
1Jan44		34	7	H.Gr.Mitte
1Feb44	39	38	1	
1Mar44		35	6	
1Apr44		42	2	H.Gr.Mitte
1May44		39	9	H.Gr.Mitte
1Jun44		39	9	H.Gr.Mitte

The **Panzerjaeger "Nashorn"** left behind by **s.Pz.Jg.Abt.519** were used to form a new **Panzerjaeger-Kompanie** for **Panzerjaeger-Abteilung 664**. As reported by the **Pz.Jg.Abt.664** on 1 September 1944: After the arrival of an **Sf. (Nashorn) Kompanie** on 18 August 1944, the **Abteilung** now has one **Sf.Kp.** and two **mot.Z Kp.** instead of three weak **mot.Z. Kompanien**. The unit is three **Jaeger** short of the authorized strength, of which one is expected to be sent after being repaired at a home base installation.

s.Pz.Jg.Abt.664 Operational Status Reports				
Date	Avail	Oper	Repair	Location
1Sep44	11	10	1	H.Gr.Mitte
1Oct44	12	11	1	
1Nov44	9	8	1	
1Dec44	9	9	0	
1Jan45	9	9	0	
1Feb45	8	3	5	
1Mar45	8	5	3	

schwere Panzerjaeger-Abteilung 88 (Hornisse)

On 3 December 1943, **s.Pz.Jg.Abt.88 (Hornisse)** with the **1.**, **2.**, and **3.Kp.** were ordered to be converted from **Pz.Jg.Abt.88** and reorganized as:

o **Abteilung Stab** organized in accordance with **K.St.N. 1106b** dated 30Mar43

o **Stabs-Kompanie** organized in accordance with **K.St.N. 1155b** dated 1Nov43

o three **Kompanien** organized in accordance with **K.St.N. 1148b** dated 1Nov43

They were issued 45 **Hornissen**, with 15 sent from the **Heereszeugamt** on 9Dec43, 11 on 19Dec43, 3 on 6Jan44, and the last 16 on 14Jan44. The unit in Mielau reported having 45 operational **Hornissen** on 31 January and were supposed to be combat ready by 10 February 1944.

Already on their way to the front on 16 February 1944, to **s.Pz.Jg.Abt.88 (Hornisse)** was ordered to be sent to Krasni and assigned to the **4.Panzer-Armee** under **Heeres Gruppe Sued**. Later transferred to **Heeres Gruppe Nordukraine**, **s.Pz.Jg.Abt.88** remained with their **Nashorn** on the Eastern Front to the end of the war. To replace losses, they were sent 10 replacement **Nashorn** from the **Heereszeugamt** on 17 June, 10 on 9 August, 10 on 26 August, 10 on 1 September 1944 and the last 4 replacements on 11 March 1945.

s.Pz.Jg.Abt.88
Operational Status Reports

Date	Avail	Oper	Repair	Location
31Jan44	45	45	0	Mielau
29Feb44	45	42	3	
1May44		10	8	H.Gr.NU
1Jun44	18	17	1	H.Gr.NU
1Sep44		19	4	H.Gr.A
1Oct44	48	42	6	LIX A.K.
1Nov44	48	47	1	LIX A.K.
1Dec44	48	46	2	LIX A.K.
1Jan45	48	47	1	LIX A.K.
1Feb45	40	24	16	XI A.K.
1Mar45	32	22	10	XVII A.K.
15Mar45	34	21	13	

Having been issued 10 **Nashorn** (sent from the **Heereszeugamt** on 9Nov44), on 28 November 1944 the **1.Kp./Pz.Jg.Abt.525** that had been sent back to Mielau was ordered to be sent immediately to Minden to the **II./Pz.Rgt.2** under **Ob. West**, to be used to build up the **Kampfgruppe Fuehrer-Begleit-Brigade**. On 21 February 1945, the available **Nashorn** with **s.Pz.Jg.Kp. (Nashorn) 93** and **525** were to be concentrated in **s.Pz.Jg.Kp.93** and this company attached to a **Panzer** or **Panzergreadier-Division**. **S.Pz.Jg.Kp.525** was to be temporarily attached to **s.Pz.Jg.Abt.654** and take over **Jagdpanther** that were still on the way.

1.Kp./s.Pz.Jg.Abt.525
Operational Status Reports

Date	Avail	Oper	Repair	Location
30Dec44	10	7	3	H.Gr.Oberheim
15Jan45	10	10	0	
25Jan45	10	9	1	
5Feb45	5	2	3	
15Mar45	5	2	3	H.Gr.Oberheim

Having been issued 12 **Nashorn** (sent from the **Heereszeugamt** on 1Dec44), the **1.Kompanie/schwere Panzerjaeger-Abteilung 93** was ordered on 28 November 1944 to be sent immediately to Minden for **II./Pz.Rgt.2** under **Ob. West**, to be used to build up the **Kampfgruppe Fuehrer-Begleit-Brigade**. On 21 February 1945, the available **Nashorn** with **s.Pz.Jg.Kp. (Nashorn) 93** and **525** were to be concentrated in **s.Pz.Jg.Kp.93** and this company attached to a **Panzer** or **Panzergreadier-Division**.

1.Kp./s.Pz.Jg.Abt.93
Operational Status Reports

Date	Avail	Oper	Repair	Location
30Dec44	12	8	4	H.Gr.Oberheim
31Jan45	8	4	4	s.Pz.Jg.Abt.654
5Feb45	7	6	1	
15Mar45	4	3	1	

Having been left behind at the front with the remaining operational **Nashorn** when the rest of the **Abteilung** was sent back to refit, the **3.Kp./Pz.Jg.Abt. 655** was renamed **s.H.Pz.Jg.Kp.669** on 16 December 1944. Four **Nashorn** were sent as replacements from the **Heereszeugamt** to **s.Pz.Jg.Kp.669** on 13 January and another 13 **Nashorn** on 10 February 1945.

3.Kp./s.Pz.Jg.Abt.655
renamed Pz.Jg.Kp.669
Operational Status Reports

Date	Avail	Oper	Repair	Location
25Nov44		12	2	H.Gr.A
1Dec44		19	2	H.Pi.Brig.70
30Dec44		17	4	H.Gr.A
25Jan45		17	4	H.Gr.A
15Mar45	13	10	3	H.Gr.A

Tactics

The **Bedienungs und Scheissanleitung fuer die 8,8 cm Pak 43/1 (L/71) Hornisse** dated 15 May 1943 was intended for **Hornisse** outfitted with the **Zieleinrichtung 37 mit Sfl.Z.F.1a**. Some of the specific information did not apply to the **1.Serie** of 50 **Hornisse** that had been outfitted with the **Zeileinrictung 34**.

1. As a heavy anti-tank weapon with the greatest penetration ability and longest range, the **8.8 cm Pak 43/1 (Hornisse)** has the ability to engage resisting targets already at long range.

It serves to combat armored targets with **Panzergranaten** and **Hohlgranaten** (shaped charge) and troops with **Sprenggranaten** (high explosive rounds) by direct fire. Indirect fire of **Sprenggranaten** from concealed firing positions is to be used only in exceptional cases.

2. The main task of the **8.8 cm Pak 43/1 (Hornisse)** is to destroy enemy tanks. In a firefight with enemy tanks, aimed single shots at favorable ranges and low striking angle are the keys to success. In addition, the **Hornisse** can be employed with good effect in combating bunkers and troops at long range.

9. The gun has two types of sights, the **Zieleinrichtung 37 mit Sfl.Z.F.1a** for direct fire with **Panzer-, Hohl- and Sprenggranaten** and the **Aushilfsrichtmittel 38** for indirect fire with **Sprenggranaten**. The range drum of the **Zieleinrichtung 37** has five cylinders with range marks in the following order: **Pzgr.39-1** graduated to 4000 m, **Sprgr. L/4.7** to 5400 m, **Gr. 39 HL** to 3000 m, **Pzgr.40/43** to 4000 m, and a **Strichteilung** graduated from 0 - 100 **Strich** (mils)

12. In combat with tanks, the correct choice of ammunition, the range, and trajectory are of decisive importance for successfully hitting the target. Firing the first shot at a specific area on a tank is only possible at close and exactly known ranges. When the range is not exactly known, the shot pattern must first be observed before precise aiming can occur. Firing the first shot with the aiming point **"Beschussflaeche aufsitzend"** (at a specific area) is allowed only at close and known ranges.

13. The following types of ammunition are to be used: **Panzergranate 39-1** at ranges up to 4000 meters against all tanks

Hartkerngranate (tungsten core) **Pzgr. 40/43** at ranges up to 2000 meters against tanks with thicker armor that are difficult to combat. It has the highest penetration ability at short range, but this falls off rapidly. Use very sparingly! Ammunition is in short supply!

Hohlgranate Gr.39 Hl (hollow charge) at ranges up to 2000 meters against all ranks. Always use in preference to **Pzgr.39-1** when it can hit (less accurate than **Pzgr.39-1** at longer ranges) and penetrate.

Sprenggranate, Sprgr. L/4.7 at ranges up to 5400 meters. Use **ohne Verzoegerung** (without delay fuze) against troops, nests of resistance, guns, and massed targets. When used against tanks it can damage the tank only by firing at weapons, gun mantle, vision slits, and sights but can destroy the tank with favorable hits on the engine vents on the rear that set the tank on fire, under the turret overhang on T 34 and Infantry Mark II, and under the turret rear on T 34. **Sprenggranaten** should be used **mit Verzoegerung** (delay fuze set) against targets behind cover (wood bunkers and houses) and by ricochet against troops (exceptional).

14. Immediately fire for effect with **Panzergranaten 39 and 40/43** at all ranges up to 2000 meters. To succeed in hitting the target when firing **Pzgr.39** at ranges over 2000 meters, first the range must be exactly determined by using a rangefinder or bracketing the target with **Sprgr**.

15. Immediately fire for effect with **Hl-Granaten** up to a maximum range of 2000 meters. At longer ranges the probability of hitting the target is too low.

16. Immediately fire for effect with **Sprenggranaten** at ranges up to 1200 meters. Bracketing to range on the target must first be used at ranges above 1200 meters.

When the **schwere Panzerjaeger-Abteilung 519** was assigned to them in December 1943, Panzer-Armee 3 introduced this new weapon by distributing the following **Merkblatt** (instruction sheet) down to battalion level.

Panzerjaeger "Hornisse"

1. The **schwere. Panzerjaeger-Abteilung "Hornisse"** is the mobile defense for the command to concentrate against strong enemy tank attacks. Employ as an **Abteilung** or **Kompanie**. Distributing under **Kompanie** strength reduces operational readiness because then control by radio, resupplying ammunition, and maintenance are not possible.

2. The **"Hornisse"** is an **8.8 cm Pak auf Selbstfahrlafette (Panzerfahrgestell IV)**. Its strengths are long-range rapid fire with accurate sights that can engage all types of tanks at ranges up to 4 kilometers and mobility as a self-propelled gun. Its weaknesses are the high superstructure and light armor.

3. The **"Hornissen"** await the enemy tank attack in ambush and firing positions. Thorough scouting is needed.

4. **Schwere Panzerjaeger "Hornissen"** are not **Sturmgeschuetz** or **Panzer** and are not to be employed like them.

5. Listen to the commander of the **"Hornissen"** before giving orders. The value of these weapons lie in the range of the gun. Their effective range is from 1000 to 3000 meters.

Operational Experience

On 10 May 1943, a status report about the schwere **Panzerjaeger-Abteilung 560 (Hornissen)** was written and sent to the **General Inspekteur der Panzertruppen** as follows:

The **Abteilung** with three **Kompanien** has been in its present shelter in Jushnyi for about 9 days, the **Stabs-Kompanie** for 2 days.

*Formation of the **Stab** (headquarters) began on 19 April 1943 in Spremberg and was completed on 1 May, after which it was transported to the East to Jushnyi. Training didn't commence until 2 days ago after it arrived in the current area. The **Stabskompanie** consists of 33% old people who were previously used only in rearward service. Its training is satisfactory. Another 33% of the troops are young replacements without any combat experience but have been sufficiently trained. The rest of the troops are for the most part **U.K.** troops which can be referred to as adequate. The **Abteilungsstab** has been completely outfitted with personnel and equipment. The vehicles are all new and for the most part still not run in. The signals equipment is in order, but training on the signals personnel only started on 6 May. The **Abteilungsstab** possesses three **Hornissen**.*

*The **1.Kompanie** began formation in Wuensdorf with **Panzer-Lehr-Regiment** on 9 February 1943 and completed on 10 March. The **Kompanie** was transferred to France on 29 March. It is outfitted 100% with personnel and equipment except for the **Hornissen**. Each **Kompanie** has 10 **Hornissen** so that the entire **Abteilung** including the **Stabskompanie** has a total of 33. As reported by the **Kommandeur** of this **Abteilung**, additional **Hornissen** will not be issued to this **Abteilung** for now.*

*The **2.Kompanie** began formation on 14 February 1943 in Spremberg and was completed on 24 March, and the **Kompanie** was transferred to France on 24 March. The **3.Kompanie** began to be formed in Wuensdorf on 8 March 1943, and was completed and the unit transferred to France on 1 April. Both of these **Kompanien** are fully outfitted with personnel and equipment.*

*The three **Kompanien** sent to France could only pursue relatively little training. A lot of time had to be spent on earthworks (construction of bomb shelters and digging in ammunition) so that only half of the stay in France was available for training. Driver training couldn't occur because fuel wasn't issued. Even though 10 rounds per gun were available for each **Kompanie**, the three **Kompanie-Chefs** reported that they couldn't be used for training the **Richtschuetzen** (gunners), **Ladeschuetzen** (loaders), or **Jaegerfuehrer** (commanders).*

*The **Abteilung** is located in favorable training grounds in Jushnyi. A very good firing range is about 2 to 3 kilometers from the shelter. Training of the **Abteilung** began immediately after the **Kompanien** arrived and is being supervised by the **XXIV.Panzer-Korps**. The Kommandierende General (General der Panzertruppe Nehring) very strongly looked after the **Abteilung** himself and from all sides support is provided to the **Abteilung**. **Heeresgruppe Sued** had immediately issued 50,000 liters of gasoline for driver training. Weapons training is also being conducted at this time. Of the 20 rounds per gun issued for training from the first allotment, 10 rounds have been fired. The other 10 rounds are to be fired in training this week. The gun crews have been sufficiently trained, but shooting ability is weak. But it can be supposed that this will be improved in a short time through intensive training. It appears to me that there exists a certain reticence to use the **Rundblickfernrohr** (telescopic sight for indirect fire).*

*The ammunition situation hasn't been settled. The **Abteilung** was issued their first allotment of 400 rounds per gun in France. But only 200 rounds per gun were brought with them to the East and, as ordered by the **1.Panzer-Division**, the rest were left with a guard unit in the West. This guard unit arrived here on 9 May, after being ordered to turn over the ammunition to the **1.Panzer-Division**. As reported by **Gen.Qu.** (head of supply in Berlin), replacement ammunition can't be sent from the home country in the near future. Therefore, the **Abteilung** has only 180 rounds per gun if they are sent into action in the foreseeable future.*

The following experience report from the **schwere Panzerjaegerabteilung 560** was written on 9 May 1943:

*The **Abteilung** has had the following experience:*
A. Personnel

*A parent unit for an **Abteilung** was not available. It was put together from all the regions in the **Grossdeutschen Reich**. About 40% of the troops have experience at the front, 20 % of them in combat. Of the 33 **Fahrer** (drivers) for the **Hornissen**, only four are old **Panzerfahrer** who have been in combat. The other 29 were trained only on **Panzer II** and **III**, and 50% of these insufficiently for a short time.*

60 % of the drivers for the wheeled and half-tracked vehicles have insufficient driving skill. They had to be sent to driving schools.

***Funker** (radio operators) were quickly trained. Their knowledge of radio procedures is good, but knowledge of the equipment is lacking.*

*About 80% of the **Waffenmeistergehilfen** (armorer assistants) were trained on **Pak** (anti-tank guns). The **Waffenmeister** (master armorer) for the **Stab** and the **Waffen-Uffz** (weapons NCO) came from the **Infanterie**.*

*The **Panzerschlosser**, **Getriebeschlosser**, and **Panzerwarte** (tank and transmission mechanics) received only a two day training course at Alkett.*

*For the **Fuehrer-Hornissen** in **Abteilung Stab** and one **Hornisse** for each **Kompanie-Chef** only one **Funker** is assigned for both radio sets (**Funkgeraet f** and **Fu 8**). Because the **Funker** for the **Funkgeraet f** sits up front and must operate this radio set himself, there isn't any crew member available for the **Fu 8**. Two **Funker** are needed for each **Fu 8**, because these transmit mainly by telegraph key. Two additional **Funker** must be assigned to each **Gruppen Fuehrer** in the **Kompanie** and the **Hornissen-Zug** in the **Abteilung Stab**.*

B. Equipment - Breakdowns that have occurred:

1. Hornissen

Within 8 days in one unit back home, five engines had to be replaced because of damage, partially from overheating.

A gear broke in the right final drive of one *Hornisse*, and a new one installed.

The main drive shafts broke in three *Hornissen* (one material failure, two because brakes were applied too strongly). Brakes adjusted.

Engine damage in one *Hornisse* (probably due to overheating, a connecting rod broke) and engine replaced.

In one *Hornisse* the engine was replaced (after overheating, the head gasket burnt through).

Engine overheating was mainly caused by 90% of the steering brakes being too strongly set and the engines over-revved, supposedly because of defective tachometers.

60% of the drives for the odometers and tachometers were torn out because the cables were too short. With radio headsets on, the drivers have difficulty hearing the engine speed. Overheating was the result. Longer cables and protection by a cover are needed so that the *Funker* doesn't step on them when climbing in.

In *Hornisse (Fgst.Nr. 310059)* there were two handfuls of drill filings and a bolt with nut in the brake drum. The brake drum and steering unit were damaged, and the brake drum had to be taken out.

In *Hornisse (Fgst.Nr. 310039)* there were file and drill filings in the housing, and the engine had to be taken out.

All the hinges on the ammunition bins have been torn off. These hinges are better suited for lids on pocket watches. The ammunition is covered with dust, and every round had to be cleaned or loading stoppages would result. Stronger hinges must be installed.

The soldered points in *Panzerkasten 23a* (intercom box) constantly failed, often after only a few kilometers travel. Replacement with *Kasten 23* is desired, because no problems have occurred with these.

A large number of vacuum tubes break in the *Funksprechgeraet f*. Technical personnel state that this is due to insufficient mounting. The mount needs to be improved. Troops haven't found any solution.

<u>Proposed Modifications</u>

1. A speaking tube needs to be installed for the *Jaeger-Fuehrer* to communicate with the *Fahrer* when the radio set fails.
2. The exhaust pipe must be lengthened to the side so that the exhaust doesn't enter the fighting compartment, resulting in tiring and poisoning the crew.
3. Widen the driver's hatch so that steering brake drums fit through. Now the glacis plate has to be taken off, which costs one day's labor.

Above: Two of the first 30 Hornisse (completed by Alkett in February/March 1943) issued to s.Pz.Jg.Abt.560, have the initial production features, including twin headlights, Pz.Kpfw.III Ausf.F brake vents, and Pz.Kpfw.III Ausf.H spockets. (KHM)

4. The diameter of the forward travel lock is too large. The gun tube starts swinging. Fastening bolts with wing nuts appears to be necessary.
5. The gunner's stand is too low. Short gunners can only look through the sights by using an ammunition box. A pad for the left knee is needed, otherwise the gunner's knee will constantly hit.
6. A *Tieflader* (low boy trailer) is needed to tow broken-down vehicles.

On 15 May 1943, **schwere Panzerjaegerabteilung 560** reported on their experience in shooting the **Pak 43/1 (L/71) Selbstfahrlafette "Hornisse"**: *The following problems have occurred when firing the Pak 43/1:*
1. *The gun sights are already knocked out of alignment after a short drive. All shots go right.*
2. *In all guns the sights drift out of alignment.* **Panzergranaten** *fired using the* **Zielfernrohr** *(telescopic sight) hit to the right and short.* **Sprenggranaten** *fired using the* **Rundblickfernrohr** *(periscopic sight) hit to the right and long. The drift is directly linked to the time since the sights were adjusted.*
4. *When the vehicle is on a slope, the traverse gear is very difficult to turn.*
6. *When the sights were zeroed in the firing position, accuracy was very good. But after driving 3 kilometers the sights had again drifted out of alignment to one target width and height (1.2 m x 2 m).*

An examination of the **Pak 43/1 (Hornisse)** with **s.Pz.Jg.Abt.560** was made on 26 May 1943: *A cursory examination already revealed that the* **Marsch** *(march) and* **Gefechtszurrungen** *(combat travel locks) were deformed and allowed play of 10 or 5 mm. The supposition was immediately expressed that the inaccuracy and misalignment of the sights were due to hard blows caused by the travel locks. To support this assumptions, driving trials were conducted with three* **Hornissen** *(driven for 30 km with travel locks on and off).*

The following deficiencies were found after the test drive. The **Pakzieleinrichtung** *(telescopic sight) had moved about 2 mils up and to the side. The* **Zieleinrichtung 34** *(periscopic sight) had less movement.*

Both travel locks don't meet the established requirements. The elevation and traverse mechanisms have a dead space of up to one turn of the hand wheel.

The ammunition stowed over the exhaust pipes is heated to about 40° C. When firing at ranges of 1000 and 2000 meters, the **Panzergranate 39** *hit short and the* **Sprenggranaten** *hit too long. A trial with cold* **Sprenggranaten** *and one warmed by driving (40° C) resulted in a large difference in the range of about 250 meters.*

On 9 June 1943, **schwere Panzerjaegerabteilung 560** recorded:

Above: Four of the first 30 Hornisse (completed by Alkett in February/March 1943) issued to s.Pz.Jg.Abt.560, have the initial production features, including brackets for stowing both cleaning rods and aiming stakes on the superstructure rear, an exhaust muffler, and boxes at the rear below the panniers on both sides. (KHM)

*The **Abteilung** reported that 12 **Hornissen** (**Fgst. Nr.**310033, 310068, 310073, 310078 for the **1.Kp.**, **Fgst. Nr.**310079, 310084, 310085, 310086 for the **2.Kp.**, and **Fgst.Nr.**310087, 310091, 310092, 310094 for the **3.Kp.**) with crews arrived to complete their authorized establishment.*

*As soldiers, the crews make a good impression; however, their training is deficient. The training period was only 5 days. Another 14 days of training is needed. The **Richtschuetzen** have fired only two shots. The **Fahrer** were trained on **Panzer II** and **III** chassis and took over their **Hornissen** two days before the trip. The **Funker** don't have any training on the **Bordsprechgeraet** (intercom) or **Funksprechgeraet "f"**.*

*The **Hornissen** have the same problems as the first ones. All brakes are poorly adjusted, resulting in almost all engines overheating. Only 5 out of 12 made the three-hour drive covering 17 kilometers (train station in Nowa Bawarija to Jushnyi). Seven remained along the route and had to be repaired there; three had to be towed in.*

*A drive shaft broke in **Fahrzeug Nr. 310033**. An engine failed in **Fahrzeug Nr. 310092**. Engine temperatures exceeded 100° C in 10 out of the 12 **Hornissen** and had to pause to cool down on this short trip of 17 kilometers.*

*It is a known fact that these vehicles leave the assembly plant in a totally deficient condition. All of the vehicles must be overhauled in the unit's **Werkstatt** (repair facility).*

*Armor plates cracked in the superstructures of **Hornissen Nr. 310064** and **310084** and were cracked and welded on **Nr. 310087**. **Fahrgestell Nr. 310083** can't be steered in reverse and both the intercom and radio are defective. The scale on the gun sight in **Hornisse Nr. 310078** is incorrectly installed so that the pointer rests on 1 instead of 0.*

*Expecting problems, three officers and 12 drivers were sent to the train station where the **Hornissen** were unloaded to possibly prevent damage to the vehicles. This wasn't successful.*

On 17 June 1943, the **Pz.Offz.b.Chef.GenStdH** reported on his visit to **schwere Panzerjaeger-Abteilung 560 (Hornisse)** under **Armeeabteilung Kempf**: *The **Abteilung** is not combat ready at this time because during drives the guns are shaken, which damages the mount and the elevation mechanism. A new travel lock is needed. In their experience report, **s.Pz.Jg.Abt.560** recorded that the **Abteilung** is not combat ready because all brakes are poorly adjusted, resulting in overheated engines. From this report one can conclude that the vehicles leave the assembly plant in a totally deficient condition and must be overhauled in the troops' repair facilities.*

Based on a personal discussion with the adjutant, Oberleutnant Buttmann, the **Abteilung Feldheer** reported the following status for **Panzerjaeger-Abteilung 655 (Hornisse)** on 1 July 1943:

*The **Stab** and **Stabskompanie** with **Kompanie 611** and **670** are in Minsk in the **H.Gr.Mitte**. **Kompanie 521** is in Orel under **Pz.A.O.K.2**.*

*Of the 45 **Hornissen** issued to the **Abteilung**, up to now one **Hornisse** was damaged by a mine (Partisanen) and sent back to **H.Za.Magdeburg** (for rebuild). Seven **Hornissen** were damaged by attacking aircraft on 21 June (five burnt out and two heavily damaged due to penetrations in the radiators, fighting compartment, gun, and elevation and traversing mechanisms) and were sent back to **H.Za.Magdeburg**.*

*The **Abteilung** now has 37 **Hornissen** distributed to the units as follows: 3 in the **Stabskompanie**, 14 in **Kompanie 611**, 7 in **Kompanie 670**, and 13 in **Kompanie 521**.*

*The travel locks (forward **Rohrzurrung** and rear **Gefechtszurrung** in the fighting compartment) have been reinforced on all 37 **Hornissen**. It is requested that the forward **Rohrzurrung** be redesigned so that it can be released and swung out of the way from inside the fighting compartment.*

*The **Schildstuetzen** (armor shield braces) in the fighting compartment are being strengthened using material and fitters from the **Abteilung**.*

*A steel bracket is being installed to strengthen the upper edge of the superstructure armor using material and fitters from the **Abteilung**.*

*13 of the 37 **Hornissen** have **Zieleinrichtung 37/43 (Svo-Seitenvorhalt optisch)** with **Zielfernrohr Sf.1a (3 x 8°)**, and 24 have the **Zieleinrichtung 34** with **Rundblickfernrohr 32** or **36**. Seven of the original 21 **Hornissen** with **Zieleinrichtung 37/43** were lost due to an air strike. One **Zieleinrichtung 37/43** was dismounted from the **Hornisse** that was lost on the mine and is with the **Abteilung** near Minsk.*

*The **Abteilung** requests that the **Zieleinrichtung 37/43** with **Zielfernrohr Sf.1a.** be standardized because:*
*a. The **Zieleinrichtung 34** can only be used with the **Pak-Zielfernrohr** for direct fire with **Panzergranaten**, while the **Rundblickfernrohr** must be used for indirectly firing **Sprenggranaten**. The disadvantage is that the gunner must be trained on two devices.*
*b. The **Pak-Zielfernrohr** that belongs with **Zieleinrichtung 34** has only one main and two adjacent "**Stachel**" (inverted triangles for aiming). The disadvantage is that the choice of leading a moving target at long range is more difficult than with the **Zieleinrichtung 37/43 mit Optischem Seitenvorhalt (Svo)** with one main and six adjacent "**Stachel**".*
*c. The **Zieleinrichtung 34** is spring mounted on three points and drifts out of alignment after every shot due to play in the mount.*

*Even if standardized outfitting with **Zielfernrohr 37/43** isn't possible at this time, the mount must be strengthened, because the **Zieleinrichtung 34** is heavier than the **Zieleinrichtung 37/43**.*

*In 11 **Hornissen**, the old elevation mechanism with two mounting points has been replaced with a new elevation mechanism with four mounting points and, up to now, no problems have occurred. As authorized by **H.Gr.Mitte**, the **Abteilung** sent two fitters back to Germany on 27 June to obtain 26 additional **Richtmaschinen**. The time it will take to complete the modification on all **Hornisse** can't be exactly determined, but is estimated to take at least four weeks.*

On 27 August 1943, **Kompanie 521/Schwere Panzerjaeger-Abteilung 655** wrote the following report about its experience in action in the defensive battle near Orel: *The **Kompanie** was in action in the defensive battle east of Orel from 11 to 27 July 1943 with a combat strength of 4 officers, 28 NCOs, 188 men, 13 **Hornissen**, and 3 **Flak-Vierling** under the **XXXV.Armee Korps**. On orders from **Heeresgrueppe Mitte**, the **Kompanie** was pulled out of the Orel combat zone.*

One KW II, 19 KW I, one General Lee, 30 T 34, one T 60, five T 70, one rocket projector mounted on a tank chassis, and three trucks were destroyed and one Mark II immobilized.

*Our own total losses in vehicles were two **Hornissen**, one **Maultier** and one **Kfz.1**. One **Geschuetzfuehrer** and one **Richtschuetze** were killed, a **Geschuetzfuehrer** missing, and two officers, 6 NCOs, and 20 men wounded.*

*The **Kompanie** was employed as a self-sufficient unit attached to an **Infanterie-Korps**. Frequently **Zuege** (platoons) were attached to different divisions. But the **Kompanie** must be employed as a concentrated unit in a single division's sector for the following reasons:*
*a. With the **Hornissen** outfitted with **Funksprechgeraete f** (range of about 3 to 4 km), it isn't possible for the **Kompanie-Chef** to command his unit by radio at longer ranges.*
*b. It is hardly possible to resupply the **Kompanie** when it is scattered over a wide area.*
*c. But it is especially difficult to keep vehicles operational when they are widely separated. It is significantly more difficult to repair vehicles and deliver repair parts. With the diminished capability of the **J-Trupps** (repair section) and numerous breakdowns - even minor damage - there were seldom more than six **Jaeger** in action at the same time. These costly temporary losses have a major impact on a **Zug**, which is then forced by this situation to send single **Hornissen** into action, which is tactically irresponsible.*

*The **Infanterie** demand anti-tank protection that always attacks in front of our **HKL** (main battle line). However, in terrain without cover, due to their weak armor and the large target area presented, **Hornissen** can't usually meet this demand, and this sometimes led to disagreements with the Infanterie. On 11 and 12 July, the **Kompanie** was ordered to attack - like Panzers - about 20 to 30*

Right:
This Hornisse, completed by Alkett in March/April 1943 and issued to s.Pz.Jg.Abt.655, has been modified in the field by adding extra armor onto the gun shield on both sides. (DT)

stationary enemy tanks by advancing down a 2 kilometer flat slope that fell off toward the enemy. There wasn't any cover for concealment while driving forward against the superior enemy artillery and tank forces. Such an advance would only have led to the destruction of the **Hornissen** before it was possible to successfully engage the enemy, and therefore this action had to be refused.

The following correct tactics have proven to be successful in action: Let the opponent advance against the **Hornissen** sitting in well-camouflaged and, if possible, positions behind cover. The enemy tanks must be moving, the **Hornissen** in firing positions (not the reverse as demanded on 11 and 12 July).

The action of a **Zug** on 13 July 1943 can serve as an example. 12 KW I and four T 34 were knocked out without a single loss to ourselves by **Hornisse** in well-camouflaged positions in the **HKL** even though the enemy had the possibility to advance under cover and the enemy tank attack was supported by aircraft.

When stationary enemy tanks are employed as artillery, success can be achieved only by first scouting on foot and only then if cover is available to close with the target in order to achieve a surprise appearance, conduct a short firefight and disappear into cover. A **Kompanie** action on 23 July can serve as an example. During a dangerous penetration by enemy tanks and infantry into the flank and rear of a **Grenadier-Regiment**, the **Kompanie** pulled into a depression, and after thorough scouting on foot, took up a position and knocked out a KW I and T 34. Until a counterattack could occur, the further advance of the enemy was hindered by repeatedly appearing suddenly on the edge of the depression for a short firefight.

The basic rule and orders that Hornissen not be employed in the **HKL** and not in the main zone of effective fire from enemy artillery couldn't always be followed because of the overall situation and the enemy superiority. Very often our own overtaxed **Infanterie** requested that armored and anti-tank weapons remain of their own volition in the forward line in order to provide the **Infanterie** with a moral and actual anchor. This would hinder a dangerous and rapid enemy strike when the troops were pulling back. In every case the **8.8 cm Sprenggranaten** were effective in bringing the enemy infantry to a halt or causing them to retreat. This type of action is very time-limited and only to be used in an extreme emergency because of the low mobility of the **Hornisse** in reverse gear and its small ammunition load. But complete avoidance of this type of action in many cases would have resulted in hindering a planned withdrawal and not be answerable to our heavily engaged **Infanterie**.

As already suggested, the leader scouting on foot in advance is the decisive factor. In comparison with the successes, the small losses to the **Kompanie** are to a large part due to conducting such reconnaissance. Only in this way can the best approach march be chosen and immediately pull into the best firing positions instead of later searching for them while being observed by the enemy.

Even commanding the **Hornisse** in combat was conducted largely by the **Geschuetzfuehrer** on foot instead of from inside the **Jaeger**. The **Jaegerfuehrer** gets an earlier view of the terrain and the situation, can more exactly direct his **Hornisse**, and usually has a better position to observe fire from the side and upwind. The main prerequisite is that sight and voice contact with the **Hornisse** must be absolutely maintained at all times.

In spite of the appearance of massive Russian tank units, they usually arrived in small combat groups. Coordination within Russian tank units was obviously not good. Most of the tanks knocked out by the **Kompanie** occurred against such advances of small groups of Russian tanks. Knocking out these advance tank groups (probably sent on combat recon) almost always led to the larger unit turning away.

On 13 July, the **Kompanie** stood in battle with a concentrated unit of 25 to 30 attacking tanks. In general, the accuracy of Russian tank fire was not good. However, the cooperation between their tank, artillery, and aircraft units was good.

Initially, the **Hornissen** were ordered to be dug into fixed positions. Basically this is to be avoided because of the small traverse arc and high superstructure of the **Hornisse**. Mobile employment is the only answer.

In contrast, it proved extremely useful for the **Kompanie** to immediately dig foxholes under the **Hornisse** in every assembly area. Just crawling under the **Jaeger** doesn't provide sufficient protection against shell fragments.

A deficiency in the tactical handling of the **Kompanie** was radio contact (the **Chef-Hornisse** broke down at the beginning of the action) with other Panzer units which wasn't possible because they have **10 watt UKW** radio sets (on a different frequency band than the **Fu.Spr.Ger.f**). We should strive to standardize outfitting of both units. Contact by using motorcycle messengers is hardly achievable because of road conditions in bad weather.

It proved a disadvantage that the **Maultiere** weren't outfitted with radio sets. The small amount of ammunition stowed in a **Hornisse** makes it necessary that the **Muni-Fahrzeuge** follow every movement of the combat vehicles, which can only be directed by radio. For the **Maultiere** to come up close to the combat vehicles, they must be armored - like the **Ferdinand**.

The **8.8 cm Pak 43/41 (Sf.)** has proven to be very good. Fire can already be opened at long range. Penetrating ability was sufficient to destroy all the enemy tanks that appeared. In one case a T 34 was destroyed at a range of 4200 meters. Opening fire at these long ranges is not prudent because of the small ammunition load. Sights *drifting out of alignment after long drives and sustained firing was minimal. Crews zeroing the sights directly before the action or in the assembly area proved to be advantageous.*

The gun is very accurate at ranges up to 2000 meters. Tanks were knocked out at ranges from 150 to 3000 meters. Projectile drift to the side was still acceptable at ranges up to 4000 meters. In one unfavorable case (strong side winds), it was necessary to aim four widths ahead of the target.

*The penetrating ability of the **Pzgr.39** was dependable at all ranges so that all enemy tank types (T 34, KW I, Mark II) that appeared in our sector could be engaged and effectively destroyed. With a direct hit, the tanks showed a 3 meter high burst of flame and then burnt out. Hit in the rear at 400 meters, the engine block of a T 34 was thrown out 5 meters and the turret was knocked 15 meters away.*

*The large smoke and dust cloud in front of the muzzle brake was a problem for observing fire when shooting at long ranges. Especially in calm wind, the smoke hindered the gunner for at least 20 seconds. Such a slow rate of fire is extremely disadvantageous during a concentrated tank attack. In addition, the strong smoke gave away the position of the **Hornisse**.*

Comments from the Gen.Insp.d.Pz.Tr. on the experience report of **Kompanie 521/Panzerjaeger-Abt.655 "Hornisse"** dated 28Jul43:

*This report clearly demonstrated that the purpose of the **Hornisse**, its capabilities and resulting tactical rules have been correctly recognized and applied. The experience gained in this action are covered in the **Merkblatt 47b/36 "Richtlinien fuer Einsatz und Verwendung von Pak Sf-Kompanien** (Guidelines for tactical employment of a self-propelled anti-tank company) dated 1 July 1943. One **Kompanie** knocking out 57 tanks in 17 days is the proof and happy consequence of correctly applied tactics in action.*

*A longer range device than the **Funksprechgeraet "f"** is not available and can't be issued because of cross interference. Increased issue of **Pak Sf.** with **Fu 8** or **Fu 5** is not possible because of device and frequency availability.*

*Opening fire at longer than 2000 meters must be avoided because of the ammunition load and supply situation. Usually, many tanks could be knocked out with the ammunition load carried. It must be clear to the **Panzerjaeger** that there isn't any weapon that can dependably hit a point target in direct fire at ranges over 3000 meters. Small deviations in the ammunition, propellant, weather influences, etc. cause a significant scattering at such long ranges. Therefore a relatively large amount of ammunition must be expended to hit a point target (such as a tank) at such long combat ranges.*

Therefore the basic rule is: First open fire when there is a good chance of a hit and therefore destruction of the enemy tank can be achieved with a small expenditure of ammunition.

The following experience report is from **schweren Panzerjaeger-Abteilung 519** on combat action in the **3.Panzer-Armee** sector during the period from 19 December 1943 to 24 February 1944. During this period **s.Pz.Jg.Abt.519** claimed to have knocked out 290 enemy tanks and only lost 6 **Hornisse** as total losses (4 of which were blown up because of a shortage of recovery vehicles, otherwise they could have been recovered over night).

1. General - During the period covered by this report, the **Abteilung** was employed in the Witebsk area and attached to the right or left **Korps** to build strongpoints. The **Abteilung** and two **Kompanien** were separated and attached to four divisions. When a **Kompanie** was employed in a division sector, it was attached to the division and sent to work together with the regiments.

*The **Abteilung** was unloaded in Witebsk on 10 December and held ready there as an **Armeereserve**. Detailed scouting was conducted immediately in the entire army sector, and originally one **Kompanie** was assigned to each two division sector. As planned, scouting was pursued up to the battalion combat post by 18 December, maps prepared, and the information distributed to all companies. The **Abteilung** planned the march routes and assembly areas in every division sector. In an **Alarm** plan (to make secret radio and telephone traffic easier) all the assembly areas and firing zones in the entire army sector were identified by letters from right to left. All the **Kompanie-Chefs** and **Zugfuehrer** were given these maps. This exact scouting down to the smallest detail proved to be really valuable and was an important part of rapidly sending the **Abteilung** and **Kompanien** into action.*

2. Tactical Experience - *The rolling and cut-up terrain was ideal for attack by small tank units and overall favored the chances of achieving tactical objectives. But it presented problems for a breakthrough by massed tank units. Only one sector of the terrain that was a few kilometers wide favored a massed tank attack. Any penetrations by the tanks would be forced into specific channels. These narrow defiles received special attention for choosing assembly areas for the mobile elements of the **Abteilung** and reserves.*

*As always, the **Infanterie** requested that the **schweren Panzerabwehrwaffen** be employed close to the **HKL** and during the first day in action even ordered the **Hornissen** to remain in and directly behind the **HKL**. After this the correct tactical employment was then regulated by a specific order from the **Panzerarmee**.*

*The **Hornisse**'s strength lies in the range of its gun and its mobility. These advantages are not well known. Therefore, tactical action that promised the **Kompanie** success was often hindered from the start by unbearable* orders. As examples:

o Instead of being assigned a task, the **Kompaniechef** was given an order on where to position his **Hornissen**

o The **Hornissen** were ordered to remain in their firing positions, when during darkness they could have pulled back.

o In spite of having their own sufficient **Panzerabwehr**, **Hornissen** were ordered to take up positions that lay far apart, unnecessarily diminishing their combat strength.

o Continuously, **Hornissen** were ordered into action as **Sturmgeschuetz** and **Panzers**.

As a **Schwerpunktwaffe** (strongpoint weapon) the **Hornisse-Kompanie** must be retained by the tactical commander and employed like "**Feuerwehr**" (firefighters) for the higher command for mobile **Panzerabwehr**.

The following fundamental rules are the basis for tactical employment of a **Hornissen-Kompanie**:

The action is controlled solely by the **Kompanie-Chef** based on his assignment and his scouting the terrain. He specifies the firing and assembly areas and, based on the situation, orders alternate positions for every single **Hornisse** from which they can view the surrounding terrain. In every case the main part of the **Hornissen-Kompanie** must at all times be held in hand, ready for mobile action.

Basically, employment as **Sturmgeschuetz** and **Panzers** is to be avoided, because the Russians constantly send in anti-tank guns with their lead attacking point elements. Both of the only **Hornissen** lost by the **Abteilung** to enemy fire were lost during such an action.

3. Tactics of Russian Tanks.

A competent reconnaissance is completed before commencing a major attack. This is sure to find seams and weaknesses in our defenses. During the entire battle they never employ large operations with tactical use of strong tank forces. They always send in infantry attacks supported by tanks after heavy artillery bombardment. The Russians place great value on rolling over, wearing down and destroying our infantry and anti-tank weapons. If at first he doesn't succeed with normal attack tactics, then he infiltrates in. At first he sends in weak forces at night in thinly manned sectors of the **HKL** to create "bridgeheads", which are then continuously reinforced. If these aren't immediately cleaned out, the penetrating enemy force can no longer be confronted.

In the defensive battle around Witebsk the Russians attacked in mass with a strength of up to 50 tanks. The tank assaults usually occurred in unexpected locations with up to 20 tanks directly after lifting the preparatory barrage. Tanks carrying infantry were the "carriers" of the infantry attack and pulled the dependent infantry forward. Russian tank attacks cleverly took advantage of the terrain, especially depressions, that ran parallel to the **HKL** and cleverly employed forward observers that themselves rode in tanks. This is the basis for the close cooperation between Russian tanks and artillery. If the major attack has high losses before reaching the **HKL**, the Russians alter their tactics. His tanks become cautious and establish strongpoints from which to fire outside the range of our defensive weapons.

Most of the tanks are T 34 with some KW I, KW II, and 15.2 cm Sturmgeschuetz. The T70 is usually used as command and radio tanks. In mixed units (T 34, KW I, and Sturmgeschuetz), the heavy tanks provide covering fire while halted and usually from good covered positions. The rapidly advancing T 34 follow in stages. Usually Russian tanks fire even when moving. In this way, part of their tanks conduct effective aimed fire, while the rest of the driving tanks mainly achieve a demoralizing effect. During major tank attacks, all T 34 were outfitted with 600 liters of reserve fuel. Two 50 liter fuel tanks were on the track guards, and two 200-liter drums were fastened onto the engine compartment.

Russian recovery service was continuously busy every night to recover knocked-out and even burnt-out tanks.

During one attack, as Russian tanks approached our **HKL** they had the turrets turned to the rear and were firing back at their own lines to fool the opponent into thinking they are returning German Panzer.

4. **Hornissen** Tactics

The **Hornissen** are to be employed usually by company in a division sector. To build a strongpoint, the **Abteilungstab** with two **Kompanien** are to be immediately employed. The **Kompanie** as an intact unit should be attached to a division and employed under the direction of the **Kompanie-Chef**. They were assigned to work with the strongpoint regiments, which has worked out well.

When attached to a division, it was almost always noticed that the **Kompanie-Chef** seldom received general orders to defend the entire division sector. Mostly, a map was used to show him at which points so many **Hornissen** were to be positioned. Therefore the **Hornissen** were often unsuitably employed and unnecessarily concentrated. Neither the range nor the excellent penetrating ability of the gun were sufficiently considered, and concentrating the **Hornissen** resulted in their being unnecessarily knocked out by artillery and mortar fire. The possibility of holding **Hornisse** behind critical points and utilizing their mobility to advance into prepared positions for a surprise attack was seldom used.

The divisions seldom agreed with proposals from the **Kompanie-Chef** based on his scouting the terrain. Under pressure from general tank-nervousness, demands were more often made to immediately get into firing position, long before any enemy tanks were reported or the direction enemy tank attacks were taking was known. Thereby, the **Hornissen** were held in a specific place in difficult terrain for long periods and thus removed from any possibility of employing them concentrated at a strongpoint. When conducting these purely defensive tasks, insufficient attention was paid to necessary maintenance. At critical points where 10 or 20 tanks had attacked, which the **Hornissen** could have destroyed, the **Hornissen** were sometimes held for days longer in positions that could be occupied only at night.

Against the flexible tactics used by the Russian tank forces, using the **Hornissen** in fixed positions is basically to be avoided. It hinders quick surprise strikes and switching to other critical points. After completing their assigned task, the **Hornisssen** should be ordered to pull out of their positions and return to the assembly area. Only this will ensure that the **Hornissen** are available to the higher command to employ as a mobile **Panzerabwehrwaffe** that can be concentrated to create a strongpoint.

Employment of all the **Hornissen**, which is also continuously demanded, is also incorrect. At least one mobile **Zug** must remain in the hands of the **Kompanie-Chef**. Once when 10 to 15 enemy tanks attacked a critical point in a division sector, suddenly 8 or more **Hornissen** were ordered to the defense. In reality in this terrain sector firing positions were only available for a maximum of four **Hornissen** which could have successfully completed this assignment by themselves. As an example, out of 47 attacking tanks, 22 were knocked out within 27 minutes without a single loss to us.

Flanking attacks usually into a neighboring sector by a **Halbzug** (section) out of a firing position had worked especially well. Firstly the **Hornissen** were out of sight of the enemy tanks, and secondly they weren't in the zone of enemy artillery fire.

It is remarkable that the Russians still don't use the name *"Hornisse"*. All captured tank crews and intercepted radio transmissions speak only about *"**Ferdinand**"* and *"Panzer"* when they mean **Hornissen**.

In general the **Panzerwarndienst** (tank attack warning system) has worked without problems. Tank concentrations and attacks of individual tanks were always reported timely. The sometimes exaggerated claims about the number of attacking tanks could be corrected by questioning several warning posts. In one case 84 tanks were reported by radio. When questioned, it turned out to be a writing error - it was actually 4 to 8 tanks.

Above: This Hornisse, completed by Alkett late in 1943, has Einheitsketten (standard track links). It no longer has a slit cut in the gun shield nor does it have an armor guard for an electrical penetration in the glacis for a second headlight. The unique quick release mechanism for the front travel lock was apparently a field modification. (BA 690/209/35)

"Nashorn" formerly named "Hornisse"

Weapons Data: 8.8 cm Pak 43/1 (L/71)
- Elevation: -5, +20 degrees
- Traverse: 30 degrees (15 R & 15 L)
- Gun Sight: Pak Z.F. 3 x 8 degrees
  Sfl.Z.F.1a (after first 50)
- Graduated to: 4000 m Pzgr.

Secondary:
- 1 - 7.92 mm M.G.34
- 2 - 9 mm M.P.

Ammunition:
- 40 - 8.8 cm Pzgr. & Sprgr.
- 600 - 7.92 mm S.m.K.
- 384 - 9 mm Patr.f.MP

Crew: Commander, Gunner
Loader
Driver, Radio Operator

Communication: Fu.Spr.Ger."f" & Intercom

Measurements:
- Length, overall: 8.44 m
- Length, w/o gun: 6.20 m
- Width, overall: 2.95 m
- Height, overall: 2.94 m
- Firing Height: 2.26 m
- Wheel Base: 2.52 m
- Track Contact: 3.52 m
- Combat Loaded: 24 metric ton
- Fuel Capacity: 600 liters

Armor Protection:
- Gun Shield: 10 mm/30° (15/30 after May43)
- Superstructure Side: 10 mm/15°
- Glacis Plate: 15 mm/70°
- Hull Front: 30 mm/12°
- Hull Side: 20 mm/0°
- Hull Rear: 20 mm/21°
- Deck: 10 mm
- Belly: 15 mm

Automotive Capabilities:
- Maximum Speed: 40 km/hr
- Avg. Road Speed: 25 km/hr
  - Cross Country: 15-28 km/hr
- Range on Road: 260 km
  - Cross Country: 130 km
- Grade: 30 degrees
- Trench Crossing: 2.2 m
- Step: 0.6 m
- Fording Depth: 1.0 m
- Ground Clearance: 40 cm
- Ground Pressure: 0.85 kg/cm^2
- Power Ratio: 11.0 HP/ton
- Steering Ratio: 1.40

Automotive Components:
- Motor: Maybach HL 120 TRM
  V-12 cyl., water-cooled
  11.9 liter
  265 HP @ 2600 rpm
- Transmission: Zahnradfabrik SSG 77
  6 forward, 1 reverse
- Steering: Differential
- Drive: Front sprocket
- Roadwheels: 8x2 per side
- Tires: Rubber 470 mm dia.
- Suspension: Leaf springs
- Track: Dry pin
  Kgs 61-400/120
- Links per side: 104

Right: This Hornisse with tactical number 300, issued to a company commander, was outfitted with the additional long range (Fu 8) radio set with the associated additional antenna mounted in the superstructure at the right rear. (KHM)

8.8 cm WAFFENTRAEGER

ARDELT-RHEINMETALL WAFFENTRAEGER

At a meeting with Krupp on 19 April 1944, Oberst Woehlermann related the history of this project as follows: *The difficulty in moving heavy anti-tank guns (especially the 4500 kg **Pak 43/41 Rh** and the 1600 kg **Pak 40 Rh**) leads in numerous cases to their loss which troops are constantly complaining about. As an example, 70 **Pak 43/41** newly emplaced in the East were blown up to prevent capture without having fired a single shot, because towing vehicles weren't available. This led to the demand by Hitler that under all conditions **schwere Pak** were to be self-propelled. The **Pak 40** with **Spreizlafette** is also too heavy for a crew to tow, and their production in this form should cease shortly.*

*As an Oberleutnant in the field, Dr. Gunter Ardelt had the same experience and made a proposal to Oberst Woehlermann to mount an auxiliary engine on the carriage of the **Pak 40** to give it expedient drive. Woehlermann declined this proposal and referred to a failed trial by Boergemanns in Kummersdorf where an example had been built and tested. Then Dr. Ardelt made an offer to Oberst Woehlermann to create in his factory a full-tracked vehicle of the simplest type for mounting guns. At the beginning of 1944, Ardelt received a contract with the stipulation that this **Waffentraeger** must be suitable for universal use as a carrier for the **le.F.H.18/40**, **Pak 40** and **Pak 43**.*

In a meeting with HDL Saur on 20 January 1944, future **Pak** (anti-tank guns) were discussed, as recorded in the following note from Professor Dr. Mueller (Krupp):

*The troops are against all anti-tank guns that are heavier than 1400 kg, because they are too unmaneuverable for the crews to manhandle. A high number of **7.5 cm Pak 40** have been lost on the Eastern Front when switching positions due to the heavy weight. It is impossible to manhandle the expedient **8.8 cm Pak 43/41** on a single axle carriage from Rheinmetall. In spite of its weight one hopes for better utility from the double axle **Pak 43** because of its all-round traverse, effectiveness at long range, and easier to move with its two axles.*

*In combat the **8.8 cm Pak** expends too much ammunition and its penetration ability is not always needed; therefore smaller caliber **Pak** must be available that even today still can be employed on the defense with advantage.*

*During the weapons meeting with HDL Saur on 20Jan44, the opinion was strongly expressed that the capability of the **7.5 cm Pak** be increased. The current **7.5 cm Pak 40** penetrates 87 mm of armor, an **L/70** penetrates about 111 mm, while an **L/60** could penetrate 101 mm. However, a **7.5 cm Pak L/70** can only be designed on a single axle carriage that weighs 2800 kg or with double axles 4000 kg and a **7.5 cm Pak L/60** on the **le.F.H.18/40** carriage is estimated to weigh about 1700 kg. HDL Saur is of the opinion that this design promises to be successful.*

*Our project to mount a **7.5 cm Pak L/70** on the double axle **le.F.H.43** carriage must be pursued, but the weight can not be greater than about 2900 kg.*

*As the "**Endloesung**" (final solution), all were in favor of a motorized **Pak** on a chassis (named "**Waffentraeger**", probably to keep Wa Pruef 6 out) that as an **Einheitsfahrgestell** (standardized chassis) can carry the **10.5 cm le.F.H.**, the **7.5 cm Pak L/60**, better yet the **L/70**, and as **Endloesung** the **8.8 cm Pak L/73**. However, it is necessary to have an elevation of about 42 to 45 degrees in order for the Pak to be utilized as a field gun firing high explosive shells.*

*Apparently as instigated by **Wa Pruef 6**, Ardelt presented two 1:10 scale models. One with only a small side shield for the **8.8 cm Pak 43** weighs 10 tons, and the other with a side shield (fragment protection) without a roof weighs 13 tons. In addition to their low weight, the primary advantage of these proposals is the low firing height of about 1,500 m and the large traverse arc of 360 degrees with side shields lowered, 60 degrees when raised. The low vehicle weight is achieved by using a frame weighing only 600 kg instead of an enclosing hull. Herr Michael from Alkett is of the opinion that under these conditions it will work and pointed out another vehicle that weighs only 13 tons. Ardelt has only incorporated components that are already available from other vehicles. The ground clearance of 500 mm with the very low firing height of 1500 mm was liked very much.*

These proposals appear very promising, but a row of problems will occur during detailed design of this quickly created proposal.

*The model with the open mounted gun was not considered feasible, because the gunner must have some protection. As demonstrated by combat experience with the **Hornisse**, 10 mm armor for defeating shell fragments provides considerable protection for the crew, because almost all armor plates on **Hornisse** in action are covered with dents from fragments and deflected bullets but rarely has a penetration been achieved. This fragment protection gives the crew a secure feeling and makes calm aiming possible. The 60 degree traverse arc is too small, but all-round fire isn't viewed as absolutely necessary, but at least 180 degrees.*

*A wood model of the **Waffentraeger Ardelt** (for mounting the **7.5 cm Pak 42**, **8.8 cm Pak 43**, and **10.5 cm le.F.H.18/40** was demonstrated for Hitler on 26 January 1944. It was decided that production of the towed **8.8 cm Pak 43 (Kreuzlafette)** should be increased as rapidly as possible to meet a goal of 300 per month. But the final goal was to mount this weapon on **Waffentraeger** from Ardelt or Steyr.*

A meeting attended by representatives from

Ardelt, Krupp, Rheinmetall-Borsig, **Wa Pruef 6,** and **Wa Pruef 4** was held in Berlin on 3 February 1944 to discuss the **Waffentraeger** design as follows: At the start of the meeting Oberst Woelhermann (**Wa Pruef 4**) presented the tactical military requirements as follows:

Neither a single nor a double axle carriage is acceptable, nor is a chassis from any Panzer. What is requested is a tracked self-propelled carriage that has full terrain-crossing capability with the following specifications:

1. A special design instead of an expedient conversion of an available chassis
2. Low firing height (not over 1750 mm)
3. Elevation arc from -8 to +45 degrees, at least 42 degrees
4. Traverse preferred 360 degrees by every elevation between -8 and +42 degrees
5. Sights - The gun must be fully employable as both an anti-tank gun and artillery piece. Therefore a **Pak-Zieleinrichtung** *with* **Winkelzielfernrohr 10 x 7°** *and* **Zieleinrichtung 34** *with* **Rundblickfernrohr** *are necessary. A* **Scherenfernrohr** *is planned for the* **Geschuetzfueher** *(gun commander)*
6. If possible a machinegun is to be installed
7. Ports on the sides for using an **MP** *should be planned*
8. **Nebelkerzenwurfgeraet** *(smoke grenade discharger)*
9. Radio sets
10. Armor made out of **SM-Stahl** *(Siemens-Marteneit steel) 20 mm thick on front, steeply angled, 10 mm on sides.*
11. Range of 140 to 200 km
12. Maximum speed 35 km/hr on roads, 30 km/hr cross country
13. Maximum ground pressure 0.7 kg/cm²
14. Ground clearance 450 mm
15. Crew of 4 to 5 men.
16. Ammunition stowage, 50 rounds, 20 rounds in ready racks. Length and diameter of complete round 1250 mm and 146 mm, weighing 23 kg.
Armament - The main weapon is a **8.8 cm Kanone L/71** *utilizing the main components from the available* **Pak** *or* **8.8 cm Kw.K.43.** *The weapon with armor shield should be traversable all round and the highest possible side and elevation arc have for employment as both anti-tank and artillery. The conversion of the weapon will be given to Krupp and Rheinmetall-Borsig. Krupp is to work together with the chassis design firm Steyr, and Rheinmetall-Borsig is to work with Ardelt, with Steyr and Ardelt having lead responsibility for the design of their* **Waffentraeger** *with weapon.*

The armor for all functionally important parts is 10 mm **SM-Stahl** *proof against shell fragments. The forward main protection shield mounted at a steep angle is to be made from 20 mm* **SM-Stahl.**

The **M.G.42** *is to be usable in all directions, that is, also toward the side and rear. Mounting on the upper edge of the* **Schildpanzers** *appears to be suitable for anti-aircraft defense.*

Rheinmetall-Borsig will receive a contract for the design of a suitable 8.8 cm gun using the **Kw.K.43,** *completion of a wood model, and completion of a* **Versuchsgeschuetz** *(trial gun)*

The vehicle already being built by Ardelt is to be completed and driving and firing trials begun. Ardelt promised to complete the vehicle in several weeks. Ardelt will receive a contact for the design of the chassis, completion of a wood model, and production of 4 **Versuchsfahrzeugen** *(trial vehicles).*

The following report reveals the outcome of testing the **8.8 cm Pak 43 - Ardelt Waffentraeger** on 27 and 28 April 1944 in Hillerleben: *The purposes of the test were to determine 1) technical functioning of the weapon mounted on the vehicle, 2) the stability and movement of the vehicle while firing, and 3) driving effects on the gun carriage.*

The **8.8 cm Pak 43** *mounted on* **Waffentraeger-Versuchsfahrzeug** *was exactly the same as a production series* **Pak 43 auf Kreuzlafette** *including the pivot mount. The vehicle had already gone through a 120 km cross country test drive without incident. This first test drive was conducted by the design firm Ardeltwerk. Parts were still missing from the vehicle, including seats for the gun crew. Holders for the gunsight, cleaning brush and rods and limiting the elevation by switching off the electrical firing circuit hadn't been completed.*

The gun was sighted and 129 rounds fired on 27 April, mostly **Panzergranaten** *because these had the strongest impact on the gun. While firing in all directions and elevations, the stability and durability of the equipment and the impact of firing on the gunner attempting to follow the tracer were observed. Also, the gun was fired without the brakes on. Round 129 shot the ring out of the muzzle brake, which was lightly damaged. One man in the neighboring firing position was lightly wounded by a piece of the ring.*

The travel lock was tested by four circuits of the **Tridelbahn** *(obstacle course) at a speed of 20 km/hr and on 28 April two circuits at a speed of 25 km/hr.*

Firing was continued on 28 April with a new muzzle brake. In spite of the strong winds a determination of the accuracy was carried out by firing at 1000 meter range. The wind swung the gun tube back and forth about 3 mils. The cause was found to be the play in the traverse mechanism. In order to eliminate this play, one track was driven onto a wooden beam so that the traversing mechanism was stabilized by the gun's weight. The pattern of hits when firing in the direction of travel was very good. 10 shots hit within a zone 70 cm wide and 60 cm high.

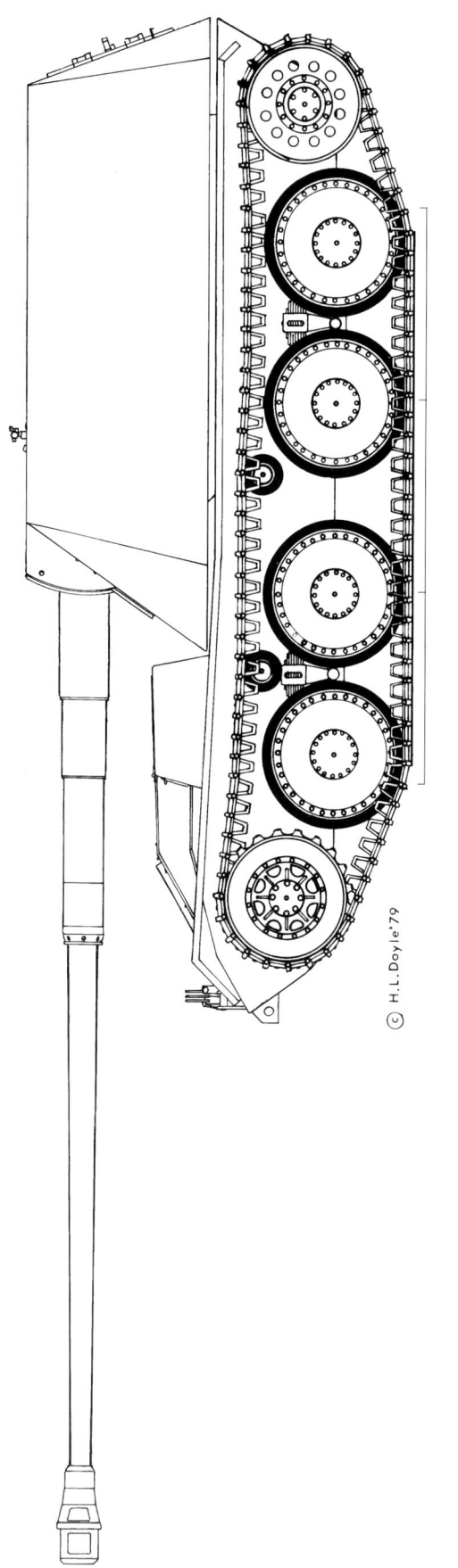

Ardelt-Rheinmetall Waffentraeger

This and Opposite Page:
This single trial Ardelt-Rheinmetall 8.8 cm Pak 43 Waffentraeger was completed by April 1944. Rheinmetall designed the gun and turret while Ardelt designed and produced the chassis utilizing Panzer 38(t) components. The specification called for the gun elevation to be as high as 45 degrees in order to use the 8.8 cm Pak 43 for long range artillery fire.
(BA-MA)

No problems occurred throughout the entire firing program, so testing was halted after 263 rounds. This was followed by a cross country drive between firing range Sued and 4000. The route went through a wide depression with steep sides and level bottom. The depression was so deep that only the gun tube and shield were above ground level and therefore presented an ideal firing position. To drive into this position, the gun was elevated or it would have bored into the ground. While driving out the vehicle was canted and the clutch in the traverse mechanism slipped, with the gun swinging to 90 degrees from the side. Due to this offsided load, the right track dug into the loose sand, causing the vehicle to cant farther over. It had to be towed out.

Test Firing Results
1. The equipment didn't show any deficiencies when firing. Stability was good in all firing directions through 360 degrees. In comparison with the other directions, stability was somewhat less when firing to the rear.
*2. The accuracy is not lower than the **8.8 cm Pak 43 (Kreuzlafette)**. The shot pattern is exceptionally good when considering the bad weather conditions.*
3. The vehicle bucked so little when firing that the tracer could be followed through the sights at ranges above 1200 meters. With an effective range is out to 2000 meters, the gunner can follow the lay of the shots by himself.
4. The gunner can maintain his eye on the sight without being hindered by shaking when firing.
5. The elevation arc is partially limited by the vehicle superstructure. The electrical firing circuit must be automatically disrupted in this zone.
*6. The traverse mechanism clutch from the **8.8 cm Pak 43** doesn't hold securely enough. Adjustment of this clutch isn't accessible as installed.*
*7. The gun equipment (stored on the **Kreuzlafette**) must be stowed on the gun or vehicle.*
8. Because the gun must be elevated to cross terrain obstacles, a special side travel lock is needed.
9. The vehicle springs are so soft and dampened that there is very little misalignment of the sight.
10. There is no ability for the commander to communicate with the driver. A signal device or a speaking tube is needed.
11. The rear tow hooks on the vehicle should be more favorably located so that the tow cable can be easily hooked on.
12. The filler opening for the fuel tank is in a poor position. The vehicle can only be refueled from cans when the gun is traversed to the side. The filler opening should be relocated to the front of the gunshield.

*Several track links were broken while driving on the **Tridelbahn** and cross country. The **RSO** track appears to be too weak for this vehicle.*

On 9 May 1944 Wa Pruef 4 sent the following report on the **Waffentraeger 8.8 cm Pak 43** to the **Reichsminister fuer Ruestung und Kriegsproduktion**: *The **1. Versuchsstueck** of the Ardelt design with **8.8 cm Pak 43 Kp** is currently being tested in Hillersleber. 277 rounds have been fired of which 210 were **Panzergranaten** without any deficiencies. The durability tests have been completed. Stability of the vehicle is very good when firing in all directions (360 degrees) and elevations.*

*Cross country and **Treidelbahn** driving trials are currently being conducted to test the carriage and travel lock as well as to determine the influence on sight alignment; followed by accuracy determination. This equipment has been driven 200 kilometers of which 130 kilometers were cross country and 25 on the **Treidelbahn** in Hillersleben. Automotive testing of the vehicle will be conducted by **Wa Pruef 6** on a **2. Versuchsgeraet** that is to be completed on 20 June 1944. The available vehicle is not yet complete. It is still missing the track guards and seats for the crew. It can be driven and fired and can be demonstrated immediately.*

On 15 June 1944 the Krupp-Berlin office informed Dr. Mueller in Essen: *The **Ardelt-Rhm. Waffentraeger** is a project that started at the same time as the **Krupp-Steyr Waffentraeger** design. This project was pushed back during the meeting of the **Panzerkommission** in Wannsee because no armor plates could be requisitioned for the **Waffentraeger** in the foreseeable future. Recently this project has been reactivated because HDL Saur is still of the opinion that the **Waffentraeger** must have **Rundumsplitterschuetz** (all round fragment protection). The **Ardelt-Rhm. Waffentraeger** has the same gun as the **Krupp-Steyr Waffentraeger** but with a new upper carriage. Firing height is at 1750 mm. It is built from **Panzer 38t** components.*

On 2 September 1944, Wa Pruef was informed that **Waffentraeger Krupp-Steyr**, **Ardelt-Krupp**, and **Ardelt-Rheinmetall** were to be demonstrated for Hitler. These three **Waffentraeger** were to be ready for shipment by 9 October. The date and location of the demonstration had not been determined.

Above and Below: This single trial Ardelt-Rheinmetall 8.8 cm Pak 43 Waffentraeger was test fired in Hillersleben in April 1944. The chassis was driven up onto timbers so that the gun was stabilized by its own weight while attempting to fire accurately in strong wind. (BA-MA)

STEYR WAFFENTRAEGER

A kickoff meeting on the design of the **Waffentraeger** was held in Steyr on 4 February 1944 attended by representatives from Steyr, Krupp, Rheinmetall-Borsig, **Wa Pruef 6**, **Wa Pruef 4**:

*Oberst Woehlermann (**Wa Pruef 4**) presented the tactical military requirements (the same exact requirements presented at Ardelt on 3Feb44 as related above). Three different weapons were discussed; an **8.8 cm Pak 43/41 Rheinmetall-Borsig**, an **8.8 cm Pak 43 Kreuzlafette Krupp**, and an **8.8 cm Pak 43/3 Krupp**. The **Rheinmetall-Borsig Pak 43/41** was dropped because series production was already running out. The **Pak 43/3** appeared to be more favorable from both Krupp designs due to the following reasons:*

*During the discussion it was determined that it wouldn't be possible to adopt an unmodified gun with unmodified carriage from either the **Pak 43** or the **Kw.K.43**. This would only have led to undesired compromises. The direction taken is much clearer that a new gun must be developed consisting of unmodified parts such as the gun tube with breech, carriage with recoil and recuperator cylinders taken over from previous designs. Far-ranging freedom must be allowed for the design of the carriage from the start in order not to burden the project with compromises.*

*With regard to the weight, a rough calculation results in a weight of 4500 kg for the **Pak 43** or the **Kw.K.43** including the **Splitterschutz** (armor body).*

*With regard to the traversable parts including armor, Krupp is convinced that the specified traverse arc of at least 70 degrees (maximum 90 degrees) can't be met with the **Pak 43** design.*

*Mounting the upper carriage from the **Pak 43** with its protruding pivot appears to be less easily adaptable than the **Kw.K.43** with a roller ring. All round fire at any elevation would be possible by adopting the **Kw.K.43**. To achieve stability for the entire vehicle the gun's recoil travel must be increased by about 700 mm to result in a recoil shock of 7.5 to 8 tons.*

*Adapting the **Kw.K.43** with jacket carriage and mantle allows the gun to be mounted at about the center of balance which will reduce stress to the elevation gear when driving. Also the shield trunnions can be mounted in the front armor with its weight utilized for the **Pak** design.*

*As far as possible Steyr intends to utilize available components for the **Waffentraeger** including the chassis, suspension, engine, transmission, and steering gear in order to complete 2 **Versuchsstuecke** (1 for **Wa Pruef 4** and 1 for **Wa Pruef 6**) with **Waffentraeageroberteil** from Krupp by the end of May.*

Wa Pruef 6** requires at least 10 horsepower per ton, and the number of cylinders in the **12 cylinder Boxermotor** can't be increased to achieve this. **Wa Pruef 6** withheld any decision on using the planned components (**RSO

Left and Opposite Page:
This wooden model for a Steyr-Waffentraeger was completed and photographed on 19 May 1944. Steyr intended to design the chassis using automotive components from their RSO. (HLD)

steering) until the end of the trials.

In order to gain time, the desired test firing of a gun on the chassis can occur during the development period by expediently mounting a *Pak 43* with modified *Kreuzlafette* on a Steyr chassis. *Wa Pruef 4* is to arrange for the timely delivery of a gun with 30 rounds of ammunition.

Wa Pruef 4 awarded contracts for Steyr to design the chassis, complete a 1:1 scale wood model, and produce 4 *Versuchsfahrzeugen* (trial vehicles). Krupp was given contracts to design a suitable 8.8 cm gun with all traversable parts for a *Waffentraeger* by utilizing the *Kw.K.43*, complete a wooden model, and produce a *Versuchsstueck* (trial gun).

The following additional details were specified during a meeting on the **Waffentraeger** in Berlin on 24 February 1944 attended by representatives from Krupp, Steyr, **Wa Pruef 6**, and **Wa Pruef 4**:

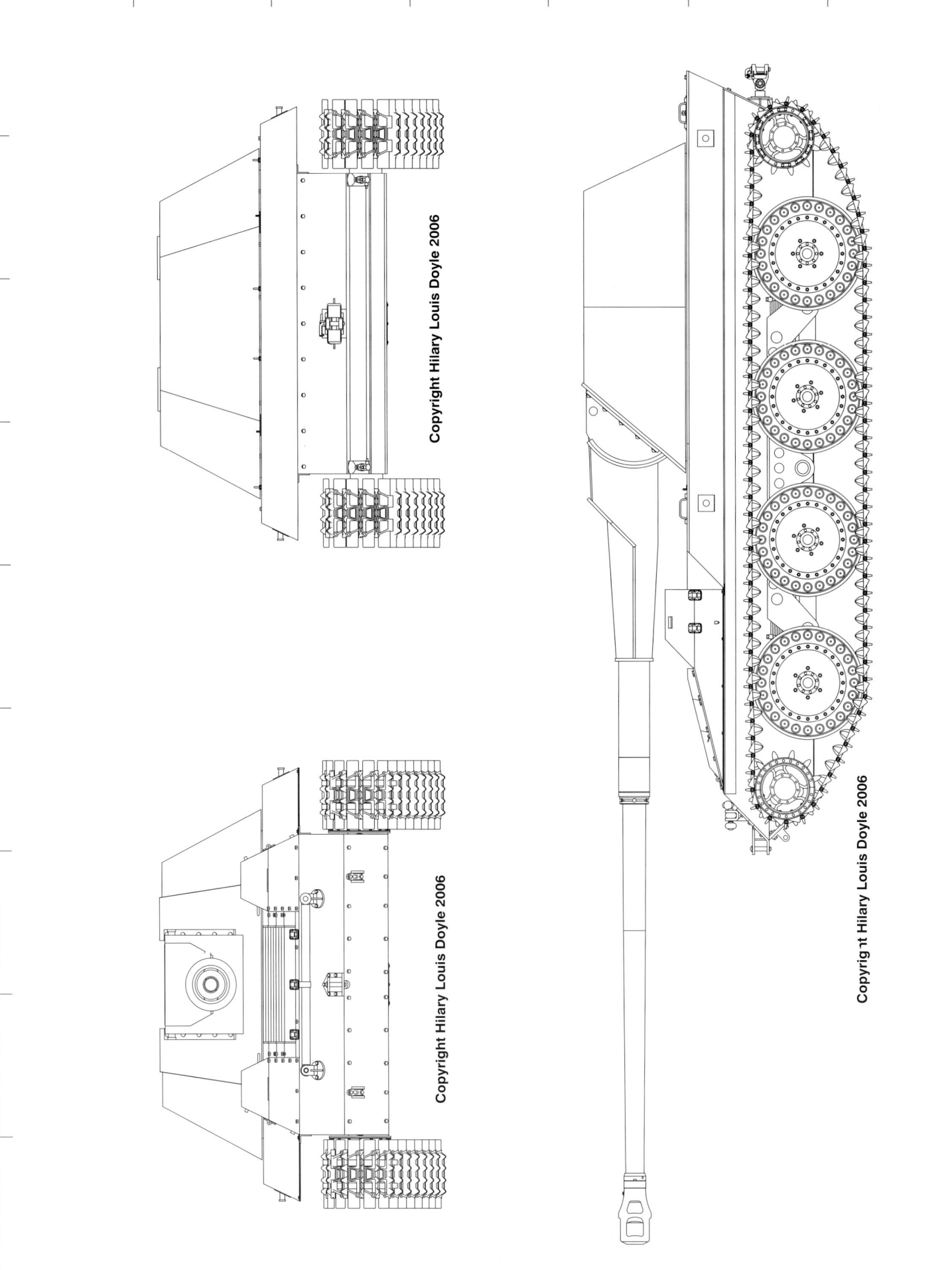

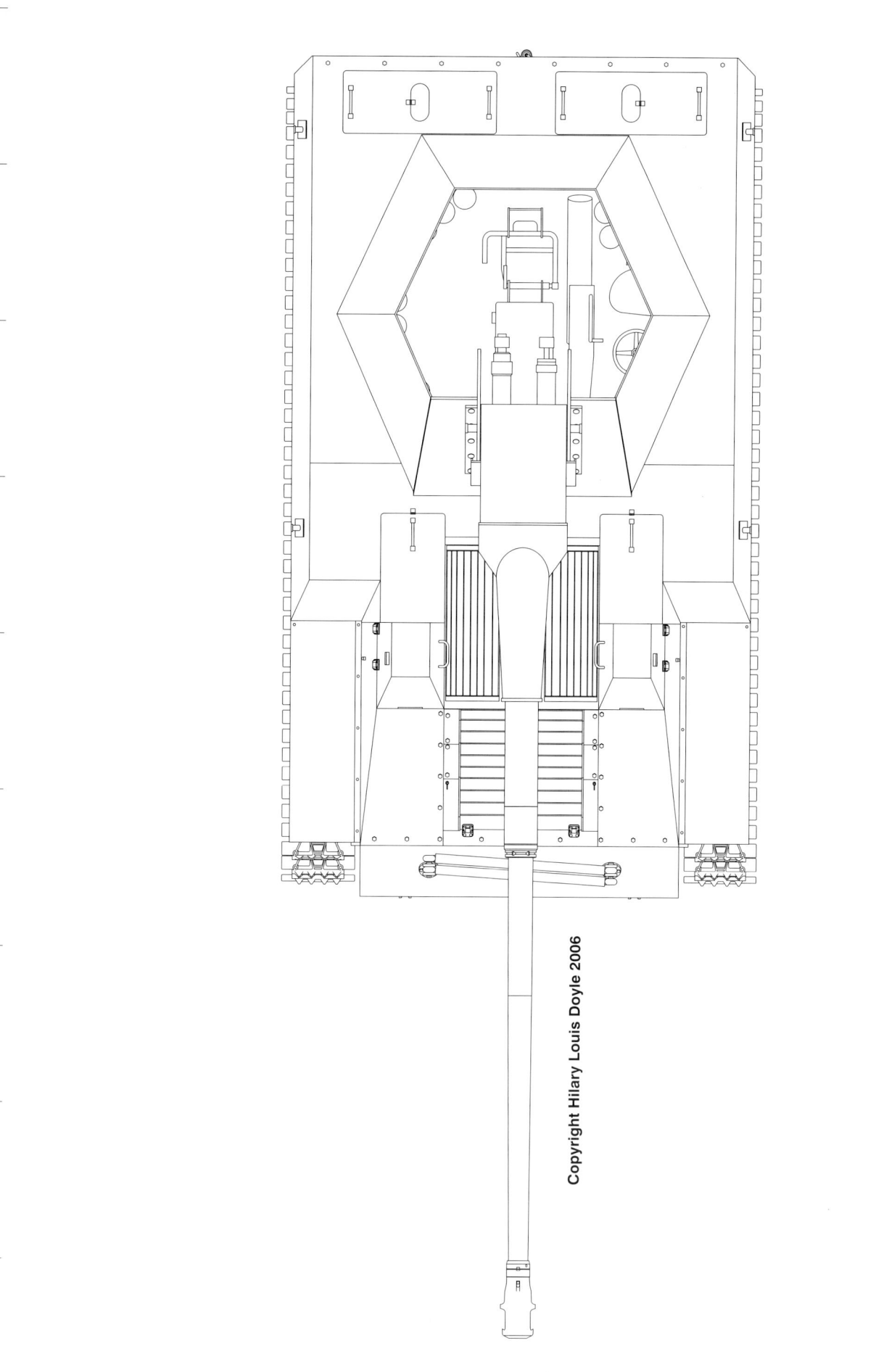

8.8 cm Waffentraeger Krupp-Steyr

I. Weapons Details

The **8.8 cm Kw.K.43** and its main components are to be used. The gun tube, trunnions, elevating mechanism, counterbalance and deflector are to be retained for the redesign. New designs are needed for the armor body, traversing mechanism, and mount for the weapon in the vehicle.

Krupp is to deliver a weapon that is transferable (to a **Reserve-Fahrzeug**) and dismountable for all-round fire. The loading and supporting devices needed for this do not need to be carried on the **Waffentraeger**. It will be decided later if the requirement for dismounting will remain as a requirement. Krupp's offer will be carried out on the first four. If dismountability isn't required later, the weight saving is about 150 kg on the carriage and about 300 kg for the four foot stands. The stands and the transfer lifting device (**David-Kran**) are to be transported on the **Munitions- (Wechsel-)Fahrzeugen**.

Krupp proposes a geared ring with inner gear teeth for the all round traversing gear and drive. Oberst Woehlermann pressed for a simple model because the capacity for large turret rings (**Tiger II**) is limited. A model made out of several segments was suggested after it was determined that outer gear teeth was just as costly as inner gear teeth. In the end Krupp proposed a chain drive that wouldn't allow continuous traversing but would have to be rotated back after completing two full turns.

This very simple design is to be specially devel-

This and Opposite Page: This trial 8.8 cm Waffentraeger had a chassis constructed by Steyr mainly using RSO components, but with a 12-cylinder engine and steel-tire roadwheels specifically designed for this Waffentraeger. The turret and gun were designed by Krupp based on their 8.8 cm Kw.K.43. (TTM)

oped. Krupp also spoke about a similarly simple clutch drive. Two of the **Versuchsgeraet** are to have a geared ring and two are to be completed with the chain drive.

Krupp and Steyr are to complete a wood model of the **Waffentraeger** with dismountable weapon by 20 March 1944 for display in the period from 1 to 10 April. This model is to be displayed at Hitler's headquarters on 20 April 1944.

II. Automotive Details

Steyr intends to use a new **12 cylinder Boxer-Motor** (140 horsepower) whose testing hasn't been completed and also the transmission, suspension, and tracks from the RSO with necessary strengthening.

Wa Pruef 6 pushed for already tested components that are available from ongoing mass production (refer to the design specifications) because there isn't any time for development work.

Wa Pruef 4 proposed to determine if the strengthened **180 horsepower Praga-Motor** for the new **leichte Sturmgeschuetz 38(t)** would be a suitable engine because large numbers are being mass produced.

Wa Pruef 6 will rapidly determine if the **8 cylinder Kloekner-Deutz-Diesel-Motor** 140 horsepower that is still in testing or another **Maybach HL-Motor** along with the associated transfer case, transmission, and steering units are appropriate for this application.

Steyr is to deliver the drawings for the planned drive and suspension by 10 March 1944 and will determine if already completed assemblies (especially the components for the **s.W.S.** from Buessing) are suitable.

Wa Pruef 6 emphasized that the steering unit for the new **Waffentraeger** can't be heavier than that in the **RSO** and doubts the utility of a simple 4-speed transmission without a transfer case for off-road.

The **RSO** tracks must be suitable for 35 km/hr speed on roads. **Wa Pruef 6** proposed that 400 mm wide **Pz.Kpfw.IV** tracks be used.

Krupp representatives made these additional notes from this same meeting on the **Steyr Waffentraeger 8.8 cm L/71** on 24 February 1944:

Wa Pruef 4 and Steyr are to receive copies of Krupp's conceptual drawing AKF 31975 of a **Steyr-Fahrgestell** with a turret mounted on rollers with space adjusted for an **8.8 cm Kw.K.43**. Elevation is -8 to +35 degrees with 360 degree traverse. A steeper angle is to be attempted for the front, if possible 50 to 55 degrees. The traverse is to have a fine drive and a fast drive to facilitate rapid coarse aiming. Dismounting the turret on pipes is not to be pursued further because they can't be carried on the vehicle due to the increased weight. However, the strong mounting ring for securing the pipes is to be retained.

Wa Pruef 4 respectfully requested that Krupp take over production of the four **Versuchsgeraete (Aufbau u. Geschuetz)**. Parts for the traverse drive can possibly be taken from the **Tiger-Programm** after the design has been settled.

Steyr wants to complete the first **Versuchsfahrzeug** in early June. Therefore the **Aufbau** with gun must be delivered to Steyr by the end of May.

On 9 May 1944, **Wa Pruef 4** reported to the **Reichsminister fuer Ruestung und Kriegsproduktion** that: *Because of enemy action, completion of the **Steyr Waffentraeger 8.8 cm Pak 43** isn't expected until early June. A deadline for the **1.Versuchsgeraet** still can't be given.*

The Krupp-Berlin office reported to Dr. Mueller in Essen on 15 June 1944: *As reported from Oberstlt. Nuernberg, this morning HDL Saur has called for all the contracted **Versuchsausfuehrung Waffentraeger** to be completed as quickly as possible. The following **Versuchsausfuehrung** are in the works: 1) **Ardelt/Krupp mit Pak 43**, 2) **Krupp/Steyr** (4 ordered), and 3) **Ardelt-Rhm. mit KwK 43** (using **Panzer 38 t** components).*

Oberstlt. Nuernberg asked for a report on when the first gun for the **Krupp/Steyr Waffentraeger** would be ready for installation. The Oberstlt. also spoke with Direktor Hacker (Steyr), who informed him that the first **Fahrgestell** is completed and for firing trials should be expediently mobilized by installing an **RSO-Motor**. This **Waffentraeger** is to be brought over to Meppen for conducting several firing trials. Dir. Hacker stated that they were striving to complete the **Boxer-Motor** for the **Krupp-Steyr Waffentraeger** in about 6 weeks.

Dir. Hacker also informed him that the wood model, especially the turret with gun, has been completed. Oberstlt. Nuernberg thought that a meeting at Steyr was needed to see the wood model and clear up unsettled questions especially about ammunition storage. Dir. Hacker gave him several photos of the **1. Waffentraeger Fahrgestell** with the wood turret model mounted. This **Waffentraeger** makes a very smart impression.

On 2 September 1944, Wa Pruef was informed that **Waffentraeger Krupp-Steyr, Ardelt-Krupp**, and **Ardelt-Rheinmetall** were to be demonstrated for Hitler. These three **Waffentraeger** were to be ready for shipment by 9 October. The date and location of the demonstration had not been determined.

ARDELT-KRUPP WAFFENTRAEGER

On 19 April 1944 Krupp representatives visited Ardeltwerke in Eberwalde to discuss the design of the **8.8 cm Pak 43 Kp auf Waffentraeger**, as follows:

*In a meeting of the **Sonderkommission I** in Berlin on 17Apr44, Oberst Woehlermann (**Wa Pruef 4**) requested Krupp to take part in the development of the **Waffentraegers 8.8 cm Pak 43** at Ardelt and to take over responsibility for the weapons design for this project.*

*The **1.Versuchsfahrzeug** for our **1.Versuchsgeraet Pak 43** made available to **Wa Pruef 4** was not completely satisfactory. Therefore, Ardelt was given a contract for a **2. Versuchsfahrzeug** with the limitation that it was only to carry the **Pak 43**. At this time Rheinmetall dropped out of participation in the development. Ardelt was sent a second **Pak 43** from series production. The **Waffentraeger** with this gun presented a significant improvement and was completely satisfactory in firing and driving trials. Generals Leed, Huether, and Schneider as well as Oberst Woehlermann were extremely complimentary of this success. HDL Saur authorized the contract for 10 vehicles of the same design with insignificant improvements that were to be completed by July 1944. Oberst Woehlermann stated that since this equipment fulfilled all requirements for maneuverability and tactical employment, it is the view of Wa Pruef 4 that the so called "**organische Loesung**" **KwK 43 Kp auf Steyr-Fahrzeug (Waffentraeger Krupp-Steyr)** is now in the background and Oberst Woehlermann has downgraded it to the 35th position on the **Dringlichkeitsliste** (priority list), while the **Waffentraeger Krupp-Ardelt** is upgraded to the 7th position.*

*The **Versuchsgeraetes 8.8 cm Waffentraeger Krupp-Ardelt** was inspected. The unaltered **Pak 43 Kp.** had been mounted on the **Ardelt-Raupenkettenfahrzeug**. It can fire all round at an elevation of 10 to 15 degrees. Toward the front it can fire from about +20 to 25 degrees down to - 8 degrees and toward the rear at about +10 to 15 degrees down to - 8 degrees. On both sides over the tracks in a sector of about 50 degrees traverse, the gun can be fired at up to +40 degrees elevation. The pivot point is somewhat behind the center of the vehicle mounted on a cross member in the hull. The firing height of the mounted gun is 1750 mm., the ground clearance 450 to 500 mm.*

*Loading is done from a platform on the vehicle. The gun tube is supported by a "scissors" travel lock on the front of the vehicle, with the arm of the travel lock about 1 meter longer than on the original **Pak 43**. The gun is absolutely secured when driving on poor roads or crossing difficult terrain. The side **Klappschild** to protect the **Richtschuetzen** (gunner) has been enlarged by Ardelt. Otherwise there isn't any additional armor other than the original **Pak 43 Schutzschild**. This allows the total weight of the **Waffentraeger** with gun to be limited to 9 tons.*

*The hull is of simple construction and basically consists of two upright side carriers that are held together to form a rectangular box by a belly plate and an upper **Riffblech** (textured plate). Unaltered tracks were taken over from an **RSO**, the suspension stems from a **Panzer 38 T**. In addition to the idler wheel, there are four roadwheels on swingarms that are paired together with a balanced leaf spring bundle. Drive is at the front. The roadwheels have steel bands instead of rubber tires.*

*The engine is a **Maybach HL 42** rated at 100 horsepower that gives the vehicle a maximum speed of 25 km/hr. The steering gears are from a **3 to Zugmaschine**, the (4-speed Kb 40 D) transmission from Zahnradfabrik. The engine is mounted forward to the right, the driver's wall forward to the left beside the engine. An upper body of the **Fahrer** (driver) is protected by an angled, hinged armor shield with a viewing slit. The area in the hull between the engine, transmission and driver's seat on the forward end and the cross member for the pivot mount is constructed as a fuel tank. The area in the hull behind the cross member has two side-by-side ammunition bins with lids for stowing about 36 rounds. A total of 42 rounds are carried, including 6 rounds stowed in an ammunition bin on the **Schutzschild**. The crew (1 **Fahrer**, 1 **Richtschuetze**, 2 **Munitioinskanoniere** and possibly 1 **Geschuetzfuehrer**) can ride on the vehicle, but there still isn't any provision for the **Munitionskanoniere** that has been selected. A second seat to the right behind the **Schutzschild** is being considered.*

*This well-proportioned **Waffentraeger** makes a completely successful impact.*

A test drive on back roads and in very difficult sandy terrain close to the factory revealed good driving qualities. It took slopes of about 35 degrees. Heavy dust clouds and soiling of the crew is to be prevented by decking over the tracks.

***Wa Pruef 4** required that the travel lock be released from the driver's seat. As it now stands, the driver must get out in front of the vehicle.*

The following was also discussed:
*1. Krupp is to send Ardelt a drawing of the enlarged left **Klappschild***
2. Krupp is to send Ardelt statistics for the change in recoil length at varying elevation
3. Krupp is to send Ardelt drawings of the new upper carriage with modified gunner's seat
*4. Krupp is to aid Ardelt in constructing 80 seamless roadwheel bands for the 10 **Versuchsgeraet***
5. Krupp is to design limiters for the traverse and elevation gear based on information from Ardelt on the highest allowable elevation while traversing through 360 degrees to prevent damaging the vehicle superstructure by a recoiling gun.

*On 9 May 1944 **Wa Pruef 4** informed the*

This Page and Opposite: The Versuchsgeraet 8.8 cm Waffentraeger Krupp-Ardelt with an unaltered Pak 43 Kp. mounted on the Ardelt-Raupenkettenfahrzeug (fully tracked chassis). (BA-MA)

Reichsminister fuer Ruestung und Kriegsproduktion: *Automotive testing of the vehicle will be conducted by Wa Pruef 6 on a 2. Versuchsgeraet (Ardelt Waffentraeger) that is to be completed on 20 June 1944.*

As described in the D 1883/2 manual dated 20 May 1944, the **8.8 cm Panzerjaegerkanone 43 (L/71)**, abbreviated **8.8 cm Pak 43**, was primarily intended to combat armored targets but because of its range (about 15,000 m) can be effectively utilized as artillery. The elevation arc was - 8 to +40 degrees with all-round traverse of 360 degrees. It had two gun sights; a **Zieleinrichtung 43 SVo** with **Zielfernrohr 3 x 8°** for direct fire and a **Aushilfsrichtmittel 38** with **Richtfernrohr 3 x 10°** for indirect fire. The center of the Z.F. was 544 mm to the left and 1.5 mm higher than the gun center. The range drums were marked and graduated for four different types of ammunition: **Gruen** (green) for **8.8 cm Pzgr.40/43** from 200 to 2400 m, **Rot** (red) for **8.8 cm Pzgr.39/43** from 200 to 4000 m, **Schwarz** (black) for **8.8 cm Sprgr.43** from 200 to 3500 m, and **Gelb** (yellow) for **8.8 cm Gr.39 Hl** from 200 to 2500 m. The gun shield protected the crew against enemy fire from the front. There was a **Visierklappe** (sighting port) in front of the gunner and a ready rack **Munitionskasten** (ammunition bin) for six **Panzergranatpatronen** mounted on the right side of the shield. The **Druckknopf** (button) for electrical firing was mounted on the handwheel for the elevation mechanism.

During a meeting on the **Waffentraeger 8.8 cm Pak 43** at the Ardeltwerken attended by **Wa Pruef 4** and Krupp representatives on 5 July 1944: *Krupp presented drawings on the modified gun for mounting on the Waffentraeger. The important points were modification of the electrical firing circuit, modification of the Schutzschild, mounting an auxiliary combat travel lock for traverse, modifying the seats, strengthening the traverse mechanism clutch, and changing the worm gear in the traverse mechanism.*

The previously hinged shield on the left side was removed. Triangular lengthening pieces are to be welded on the sloped sides of the Schutzschild on both the right and left side to protect the gunner and loader from being hit from the side. These additional shield parts are braced onto the upper carriage to reduce shaking while driving.

The new combat travel lock consists of a pin that automatically snaps into place when the gun is prepared for action. This combat travel lock can be welded to the upper carriage at any time.

The seats were redesigned and are different from the models in series production in having padding on the seat plate and back rest. In addition, the height and angle of the back rest can be adjusted, and it has a spring.

Production of the Waffentraeger Ardelt I

At a demonstration on 30 May 1944 in Kummersdorf, the decision was made to produce a **0-Serie** of 100 **Waffentraeger Ardelt I**, of which 82 would be completed with the **8.8 cm Pak 43** and 18 as **Munitions-Traeger fuer 8.8 cm Pak**.

On 31 July 1944, **Wa J Rue (WuG)** reported the schedule for **8.8 cm Pak 43 Waffentraeger** production as 16 to be completed in August, 33 in September, 33 in October for a preliminary total of 82. Including the **Munitions-Traeger**, 20 **Waffentraeger Ardelt** were to be completed in August, and 40 each in September and October. On 31 August 1944, **Wa J Rue (WuG)** reported that 0 **Waffentraeger** had been completed.

As reported on 6 October, the first 20 were expected to be delivered by the end of the month. On 16 October 1944, every means was to be utilized to complete the **0-Serie** of 100 **Waffentraeger Ardelt I (ungep.)**.

In the meeting on development and production on 12 December 1944 the status of the **Waffentraeger 8,8 cm Pak 43 (Ardelt)** was reported as: *The first 10 were to be delivered by the end of December and another 10 by 15 January 1945. It wasn't possible to meet the schedule for acquiring smithed parts and working suspension parts. Except for swingarms, material for the current series from 21 to 100 is currently being worked on or already secured. These swingarms are of the old BMM design that have to be reworked The guns will be taken from the normal production series.*

During the meeting on development and production on 9 January 1945, the status of the **Waffentraeger 8,8 cm Pak 43 (Ardelt)** was reported as: *In spite of using couriers to deliver suspension parts, only 2 are expected to be completed by 31 December 1944. All means will be used in an attempt to obtain all of the suspension parts by the end of December. If this is achieved, a further 18 should be completed by 15 January 1945.*

Operational Experience

K.St.N. 430 dated 1Nov44 for a **Batterie 8.8 cm Kanone 43 (Sf.) (6 Gesch.) (tbew./mot)** authorized this unit to be outfitted with 6 **Geschuetze 8.8 cm Kan.43 (Sf.)**

On 16 February 1945, the **Oberbefehlshaber H.Gr.Weichsel** informed the firm Ardelt, Ederswalde: *In accordance with the wishes of the Generalinspekteurs der Panzertruppen "Panzerjaeger Alarmkompanie Ederswalde" is to be raised as a field expedient by H.Gr. Weichsel. The seven completed Waffentraeger mit 8.8 cm L/71 are to be used to create this company. Oberleutnant Ardelt is assigned as the Kompaniechef. Gun crews will be provided by Heeresgruppe Weichsel. The "Panzerjaeger-Alarmkompanie Ederswalde" is to be sent into action in Ederswalde. Combat orders will be made personally by the Oberbefehlshaber H.Gr. Weichsel.*

Above and Below: A 2.Versuchsgeraet Ardelt-Waffentraeger was completed by Ardelt as shown in these photographs dated 23 November 1944. Krupp had modified the 8.8 cm Pak 43 mounting and a braced side shield was added to provide better protection for the gunner. (BA-MA)

Above and Below: Out of a contract for 100, Ardelt completed seven Waffentraeger 8,8 cm Pak 43 (Ardelt) by mid-February 1945, which were ordered to be used for the immediate defense of Ederswalde. This production series Waffentraeger 8.8 cm Pak 43 (Ardelt) is on display in the Kubinka museum. (TA)

8.8 cm PaK 43 auf Leichte Waffentraeger GW638/18 Sf

After the demonstration on 3 June 1944, the **Hoeh. Offz.f.Pz.Art.** remarked about the **Ardelt Waffentraeger**: *The advantage of this model is the use of available components and relatively quick start of production. Otherwise, all are unanimous that it is only an expedient design that was made especially for the **8.8 cm Pak 43**. HDL Saur stated that the frontline troops' desire for armor protection can in no case be considered. A working design team from Ardelt-Steyr-Krupp has been established to examine further development. The required specifications are for **Rundumfeuer** (all round fire) and **Umsetzbarkeit** (dismountability).*

This team was directed to design a **Waffentraeger** based solely on **Panzer 38t** components, as discussed in a meeting on 25 October 1944 between the **Gen.d.Artillerie** and **Waffenamt**: *In order to standardize production, the **Waffentrager** must be converted from the current design to using "38 t" components. These **Waffentraeger** with weapons and armament will weigh about 15 tons. All-round fire with 10 mm all-round armor is possible. The **Gen.d.Art.** requests a standardized **Einheits-Waffentraeger** for **le.F.H.** and **8.8 cm Kan.** with all-round fire, all-round armor, dismountable gun, and a towed carriage with easily removed holders; also usable as a **Munitions-Fahrzeug**.*

Shortly thereafter a decision was made to use components from the **Pz.38 "Reich"** (also known as the **38 D**) as discussed in the meeting on **Waffentraeger** at **Wa Pruef 4** on 13 November 1944 attended by Major Zschukke (and others from **Wa Pruf 4**), Oblt. Ardelt (Ardelt), Obering. Kracht (Auto-Union) and Koelkeskamp (Krupp):

*As directed by HDL Saur, an **Arbeitsgemeinschaft** was created to produce 550 **Waffentraeger** monthly, including **Kugelblitz**. Obering. Kracht from Auto-Union in Chemnitz is the leader of this **Arbeitsgemeinschaft**.*

*Obering. Kracht reported that HDL Saur has ordered production of the **Waffentraeger** to start with deliveries in March 1945 in accordance with the following plan: 5 March, 15 April, 30 May, 50 June, 80 July, 120 August, 170 September, 250 October, and 300 November.*

In order to meet this order, the design firms must deliver drawings to the plate manufacturers by the end of November and the drawings for working the plates by the end of December so that manufactures have sufficient time available to deliver the needed tools (boring machines, etc.).

*Then Major Zschukke reported that the following **Waffentraeger** are to be designed using components from the **Pz.38 "Reich"**:*

*1. Ardelt - Four roadwheels, unarmored for **8.8 cm Pak 43**, **s.I.G.** and **le.F.H.18/40**, with emphasis on the **Pak 43**. The **le.F.H.18/40** is only a contingency in case insufficient **Pak 43** are available and then the **le.F.H.18/40** can be mounted. Oblt. Ardelt reported that he has already started development and will strive to meet the schedule requested by Obering. Kracht.*

*Ardelt gave the weight of this **Waffentraeger mit 8.8 cm Pak 43** as about 14 to 15 tons. The **Waffentraeger** should carry 46 rounds of **Pak 43** ammunition, somewhat more **le.F.H.18/40** ammunition.*

*2. Krupp - Six roadwheels for **le.F.H.18/40** with all-round armor, **12.8 cm K.81** with shell fragment protection, and **s.F.H.18** with shell fragment protection.*

*It is of great importance that the design for the design of the **Waffentraeger** that the necessary drawings of the **Pz.38 "Reich"** be obtained from Alkett. In addition, several of the first **Pz.38 "Reich"** vehicles must be diverted for completion of **Versuchs-Waffentraeger**. Obering. Kracht thought that this would be possible. A total of about 5 or 6 would be needed.*

*Obering. Kracht clarified that the **Pz.38 "Reich"** drawings would be completed by 1 December as promised in writing by Alkett at the last meeting. As related by Hptm. Maas the total weight of an equipped **Pz.38 "Reich"** was about 16 tons.*

The first topic of the meeting of the entire **Entwicklungskommission Panzer** (led by Dr. von Heydekamp) held on 25 January 1945 was the **Waffentraeger**:

*Hptm. Evers reported: The standardization of this **Waffentraeger** for mass production is absolutely necessary due to economic reasons. The development of the **Ardelt-Krupp Waffentraeger** on the basis of 38 D components is shared by both types:*

*a. **Leichter Waffentraeger** with 4-Rollenfahrwerk, 12 Zylinder-Tatra-Diesel-Motor for **8.8 cm Pak 43**, **le.F.H.18/40**, and **s.I.G.33***

*b. **Mittlerer Waffentraeger** with 6-Rollenfahrwerk and longer frame for **s.F.H.18/40** and **12.8 cm Pak 80** (Jagdtiger-Kanone).*

*Both types of **Waffentraeger** are to have all-round shell fragment protection with the **leichten Waffentraeger** about 14 to 16 tons and the **mittleren Waffentraeger** up to 20 tons. The **mittlere Waffentraeger** will have the same engine, driver's compartment, and components as the **leichte Waffentraeger**.*

*The development of an **Einheits-Waffentraeger** with **4-Rollenfahrwerk** for 20 ton will be rapidly started and pursued by Ob.Ing. Michaels (Alkett).*

As of 25 January 1945, both the **leichte** and **schwere Einheits-Waffentraeger Krupp-Ardelt** were to be powered by a **Tatra TD 103 P** engine rated at 207 horsepower at 2250 rpm, connected to a 5-speed **AK 5-80** transmission.

From an Rheinmetall-Borsig report dated 13Feb45: *HDL Saur has requested that only Herr Michaels from Al-*

kett develop the 12.8 cm Waffentraeger using the 12.8 cm Pak 80 gun.

Plans for producing the **Waffentraeger 38 D** at Auto-Union (the sole assembly plant) were recorded in the **Panzermontageprogramm** (tank assembly program) dated 30 January 1945 as the first 10 in March, 25 in April, 50 in May, 70 in June, 100 in July, 150 in August, and increasing to 300 per month in December 1945. As related in the **Fahrzeug-Programm** dated 13Feb45, scheduled production had been delayed and decreased to plans for the first 5 **Waffentraeger 38 D** to be completed in May, 10 in June, 20 in July, 30 in August, 50 in September, 70 in October, 100 in November, and 100 in December 1945. As recorded in the **Panzer-Motoren-Programm** dated 20Feb45, Auto-Union was to complete their first 5 **Tatra 103** diesel engines in February, followed by 10 in April, 25 in May, 40 in June, 35 in July, 65 in August, 100 in September and 125 in October 1945.

The **Waffentraeger** was not discussed on 14Mar45 during a **Gen.Insp.d.Pz.Tr.** meeting on development questions after a **Notprogramm** (emergency program) had limited **38 D Fahrzeug** production to 250 per month. Based on organization factors, the **Gen.Insp.d.Pz.Tr.** decided that all of these should be completed as 225 **Jagdpanzer 38 D** and 25 **Bergepanzer 38 D**.

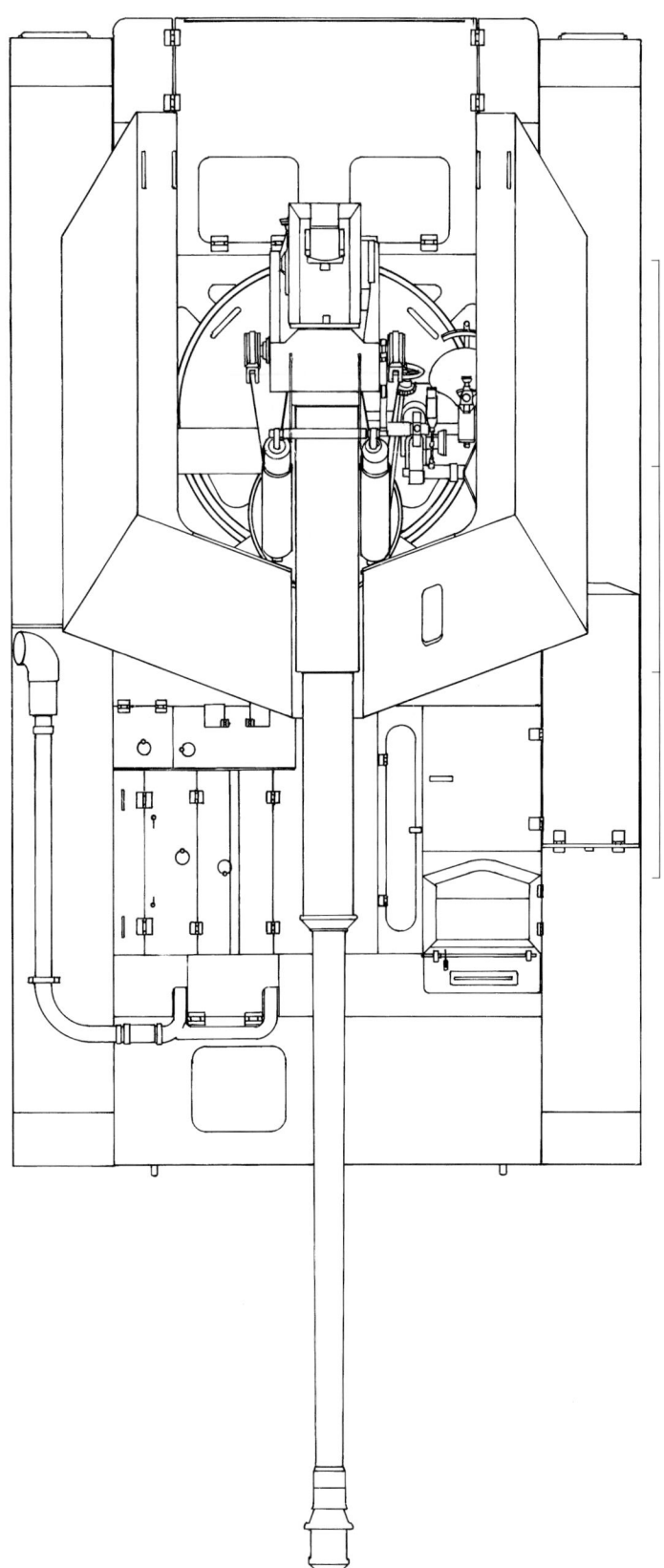

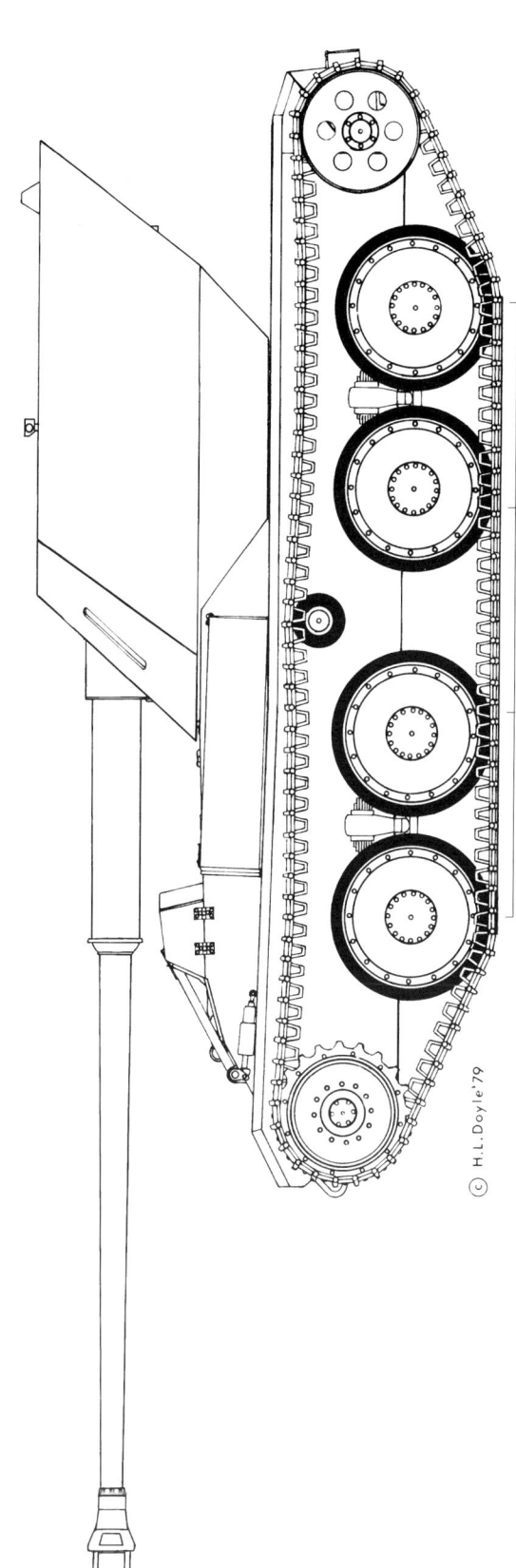

**8.8 cm Pak 43 auf
Leichtem Waffentraeger
GW 638/18 Sf
based on drawing OU 83719
dated 13Mar45**

Pz.Sfl. fuer 12.8 cm K40
previously known as Pz.Sfl.V

Starting in 1939, preliminary designs were conceived for self-propelled guns, known as **schwere Betonknacker** (heavy concrete busters), which were intended to be used against strong fortifications such as those being constructed by the French. Utilizing a previously designed **12.8 cm Flak** gun, the design of the superstructure and gun was assigned to Rheinmetall and design of the chassis was assigned to Henschel.

A **12.8 cm Kanone L/61** was to be designed utilizing the **Flak** gun tube with carriage and recoil cylinders from the **12.8 cm Flak Geraet 40**. The gun by itself weighed 7.835 metric tons. Mounted on a chassis and surrounded by an open top superstructure, traverse was limited to 12 degrees (7 degrees right and 5 degrees left) with an elevation arc from -15 to +10 degrees. A **12.8 cm Sprenggranate L/4.5 mit Rauchentwichler Nr.9** weighing 26 kg was fired at a muzzle velocity of 880 meters per second. A **12.8 cm Panzergranate mit Lichtspurhuelse Nr.4** and weighing 26.35 kg (including 490 gram HE filler) was also fired at a muzzle velocity of 880 m/s. Dials for the gun sight were graduated in units of 100 meters from 0 to 3000 meters for both the **Sprgr.** and the **Pzgr.**

Components designed for the **VK 30.01** were to be incorporated into the **Versuchs-Fahrgestell** (trial chassis) for the **Pz.Sfl.V**. Details on the components and design considerations were related in the following Henschel report on the "Sfl.V":

<u>Weight</u>: 36 metric tons, maximum speed 19.6 km/hr
<u>Pz-Wanne</u> *(armor hull)*: Special model due to its length and mounting a Rheinmetall-Borsig 12.8 cm gun that is to be dismountable. Upper wall sides are only to be 30 mm thick. Superstructure without roof. The hull is longer than the **VK 30.01** because of the location of the gun mount, but it is of single piece construction like the **VK 30.01**.
<u>Gleiskette</u> *(tracks)*: Same as the **VK 30.01**, but longer.
<u>Motor</u>: Maybach 6 cylinder **HL 116**, but a special model because the engine is higher than the radiators.
<u>Kuehlung</u> *(Cooling)*: Same as the **VK 30.01** but mounted lower. The fan drives are aligned differently, as well as the cooling air path.
<u>Ketten-Antrieb</u> *(track drive)*: Lower gear ratio because of the longer track contact requires increased steering capability.
<u>Laufraeder</u> *(roadwheels)*: One wheel more (8 total per side) because of a higher total weight than the **VK 30.01**. Oberbaurat Kniepkamp had concerns that the track would be thrown because of the great length between the last outer roadwheel and the idler wheel. The order for the inner and outer roadwheels was selected so that it results in both ends having 1300 mm distance between the track leaving the roadwheel and being supported by the drive sprocket or idler.

<u>Stabfedern</u> *(torsion bars)*: *As a backfit modification the last two roadwheels on both sides had stronger torsion bars. This was done to eliminate the vehicle rocking so strongly that the gunner couldn't observe the target again until after the projectile had already hit.*

Armor on the front consisted of 50 mm thick rolled plates, while armor plates on the hull sides were 30 mm thick. A crew of five manned this self-propelled gun.

As usual, the names applied to this project evolved with time, as follows:
o **12.8 cm Selbstfahrlafette L/61 Pz.Sfl.V**
   (Rheinmetall, 1940)
o **Schwere Betonknacker**
   (**In 6** 30Jul41)
o **Pz.Sfl. fuer 12.8 cm K.40 (Sd.Kfz...)**
   (**Wa Z 4**, 14Jan42)

In the report on renaming guns dated 14Jan42, **Wa Z 4** clarified that the **12.8 cm K.40 (Pz.Sfl.)** had been known previously as the **12.8 cm K. (Pz.Sfl.)** during **Entwicklung** (development) and the **Pz.Sfl. fuer 12.8 cm K.40 (Sd.Kfz...)** had been known previously during **Entwicklung** as the **Pz.Sfl.V**.

Production

Henschel was awarded a contract by **Wa Pruef 6** to complete two **Versuchs-Fahgestell** (trial chassis) and Rheinmetall was given a contract by **Wa Pruef 4** to complete 4 guns. It was still being designed in 1940. On 25 April 1940 **Wa J Rue (WuG 6)** reported on the status of new models in development by **Wa Pruef 6**, including a **13 cm Kan. (Pz.Sfl.) auf Fahrgestell des VK 30.01** weighing about 35 tons. At this time the two **Versuchssteueck** were planned to be delivered in December 1940. If testing proved these to be suitable, **AHA** intended to order about 100.

As reported in July 1941, the detailed drawings needed for production were available to meet the scheduled delivery of both self-propelled guns in August/September 1941. Finally, both **Pz.Sfl. fuer 12.8 cm K.40** were assembled and tested by Rheinmetall-Borsig in early 1942.

Combat Employment

By the time they were completed, the original concept of employing these self-propelled guns as "bunker busters" against the Maginot Line had been overtaken by events. A decision was then made to employ them as long-range tank destroyers. As ordered by the **Org.Abt.** on 15 May 1942, both **12.8 cm Kan. Sfl.** and the single surviving **10 cm Kan. Sfl.** were assigned to a **Panzerjaeger-Zug** that was incorporated into **Panzerjaeger-Abteilung (Sfl.) 521**.

The armor penetrating capability of APHE (armor

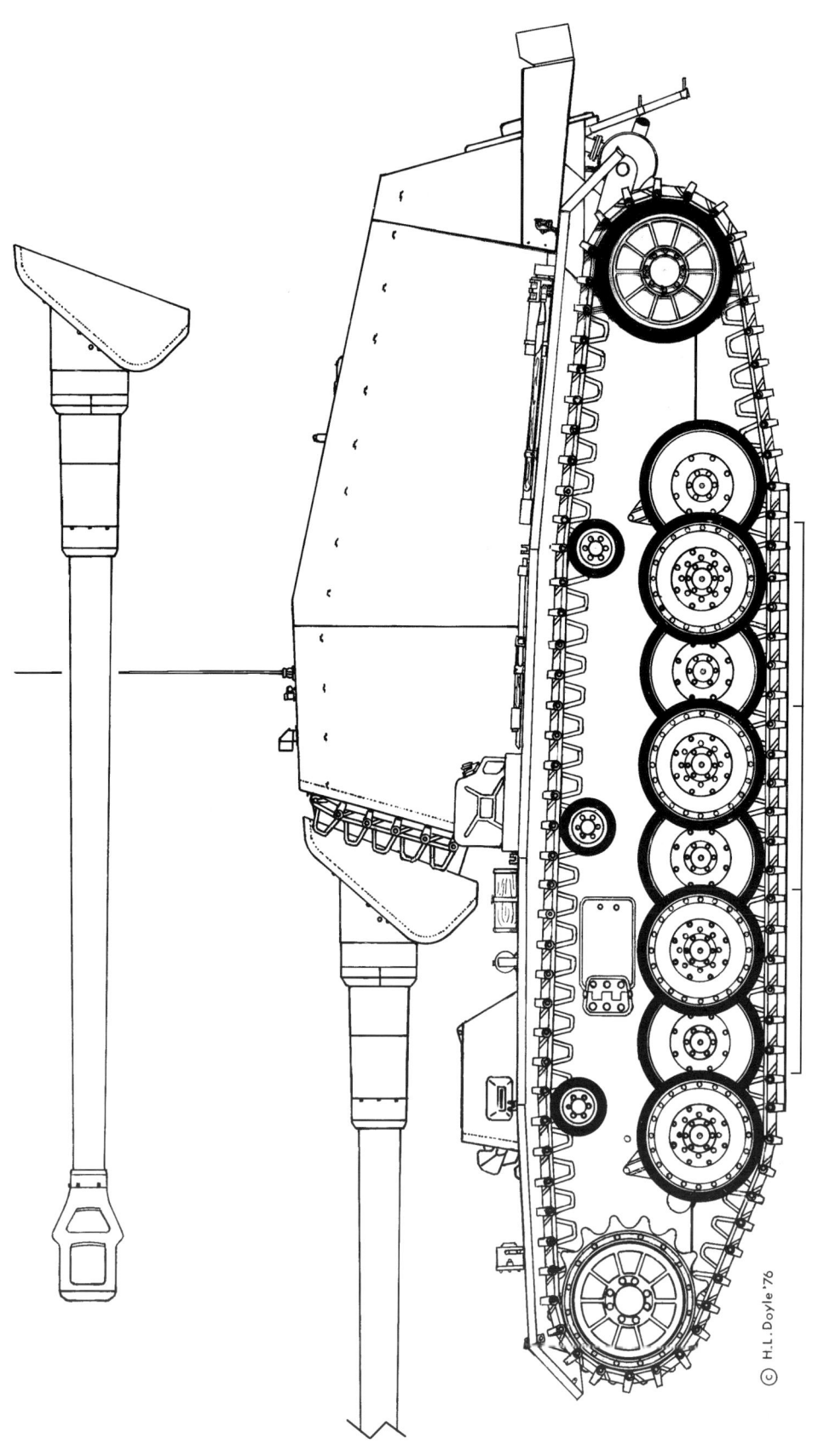

A completed Pz.Sfl. fuer 12.8 cm K.40 outside an assembly hall at Alkett on 9 March 1942. (Rhm)

A Pz.Sfl. fuer 12.8 cm K40 inside an assembly hall at Alkett. A sheet-metal compartment on the right hull front was disguised as a co-driver's compartment. (Rhm)

This Page: A Pz.Sfl. fuer 12.8 cm K.40 in an assembly hall at Alkett. A smoke grenade rack was mounted on the right rear (below) next to the crew access hatch. Opposite and Following Pages: Detailed views of the interior of a fully-equipped fighting compartment in a Pz.Sfl. fuer 12.8 cm K.40. (Rhm)

7-241

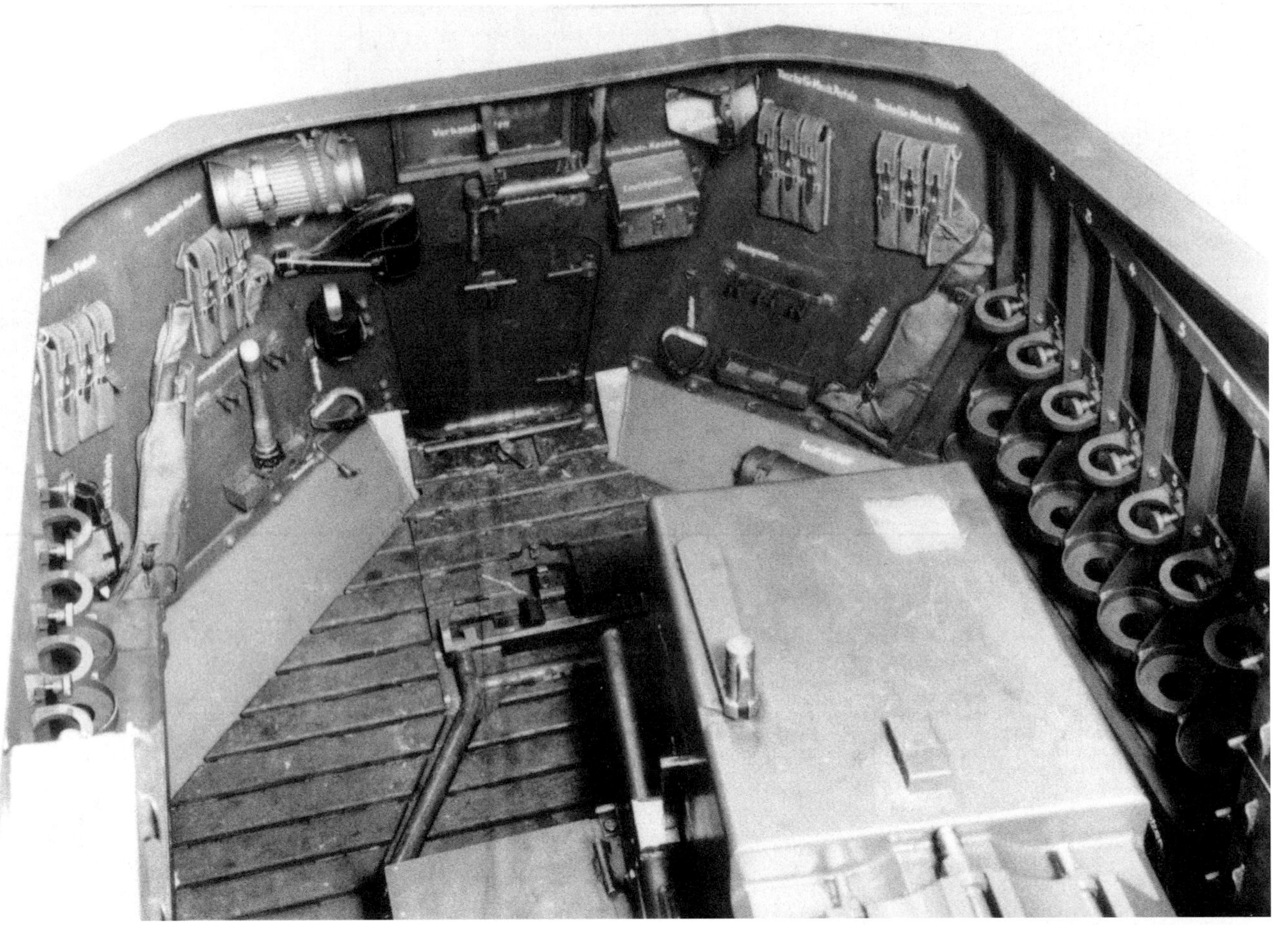

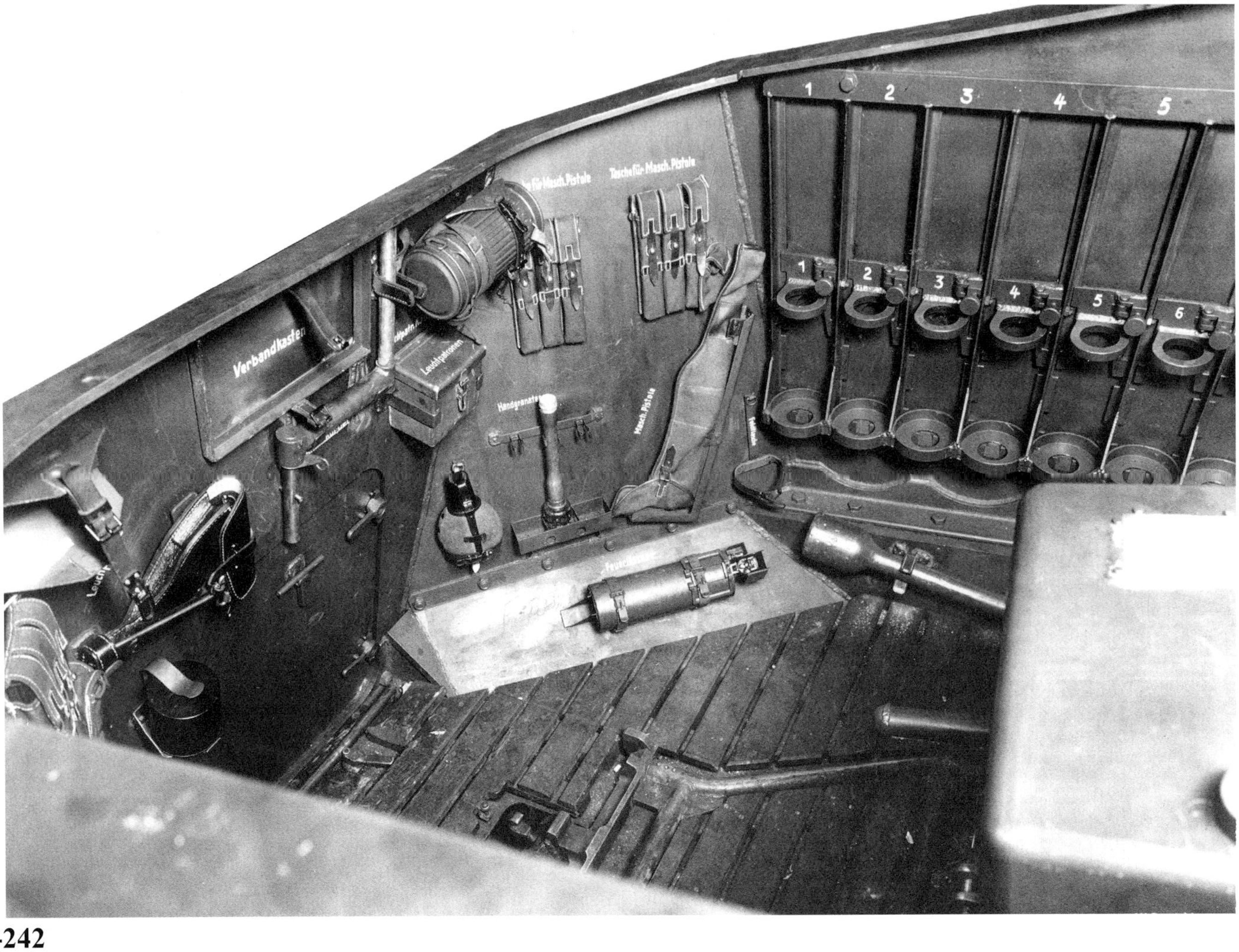
1 2 3 4 5
Tasche für Masch. Pistole
Verbandkasten
Leuchtpatronen
Handgranaten
Masch. Pistole
1 2 3 4 5 6

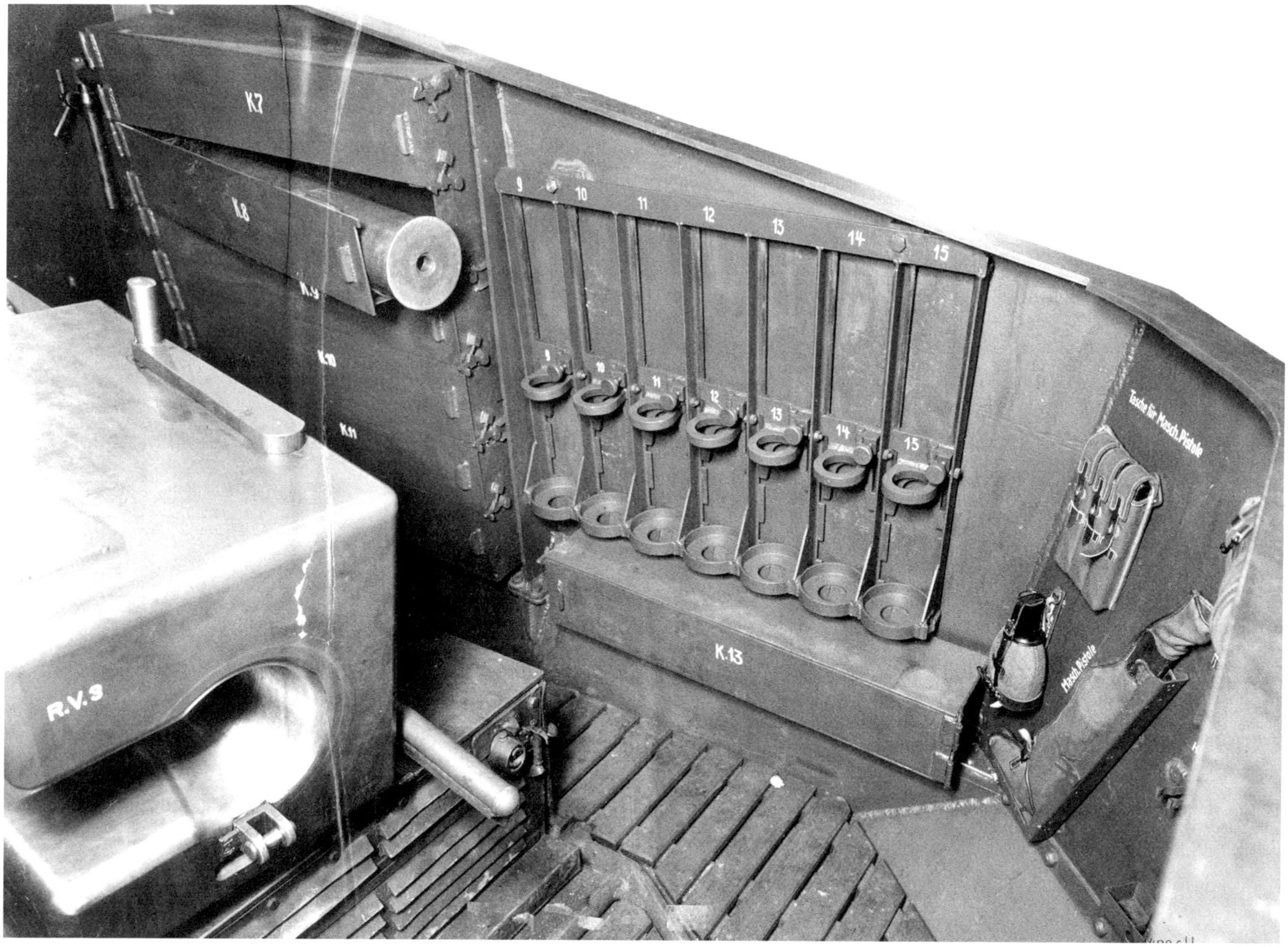
K7
K.8
9
10
11
12
13
14
15
K.13
R.V.3
Tasche für Masch.Pistole
Masch.Pistole

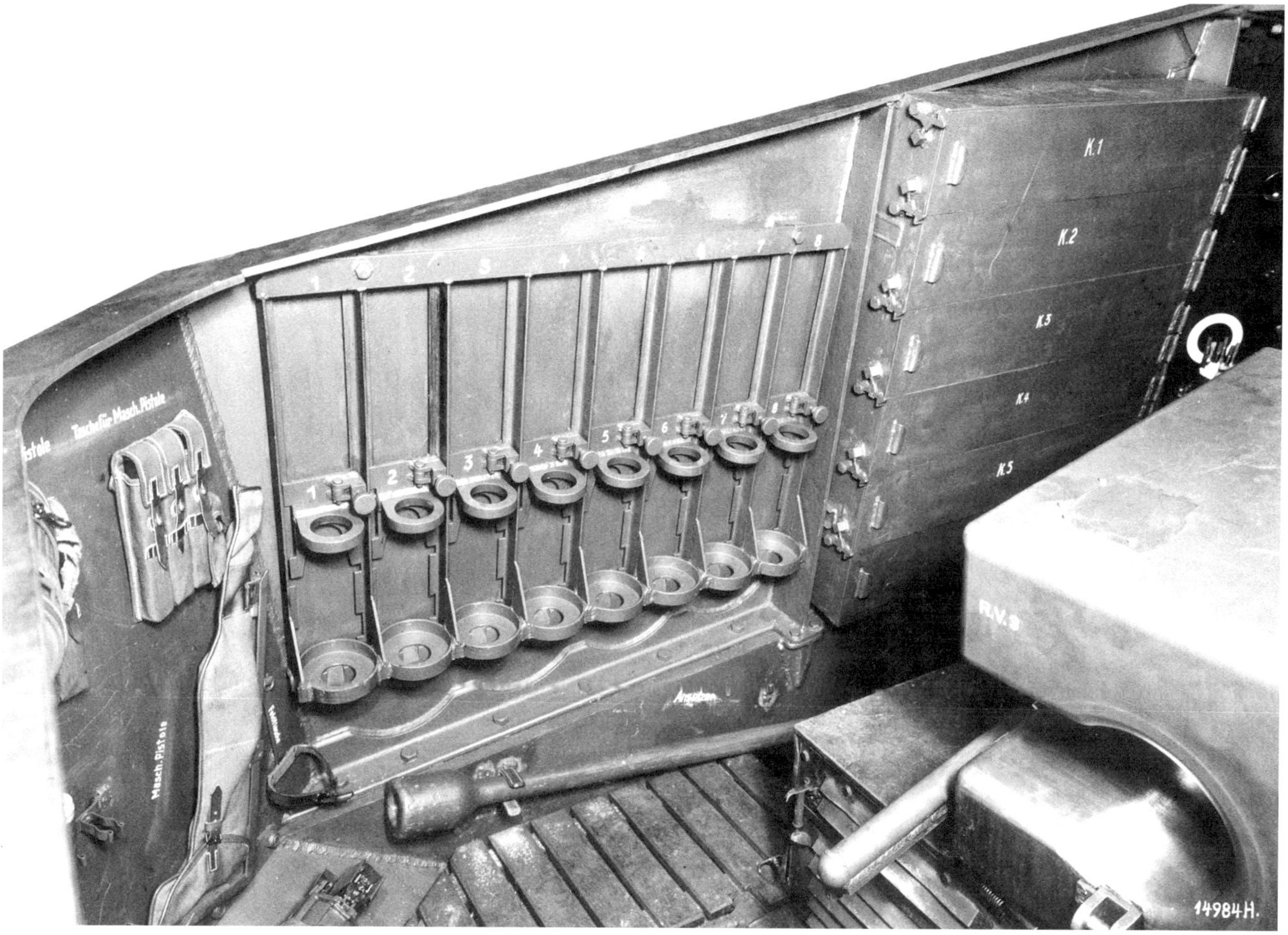
K.1
K.2
K.3
K.4
K.5
1
2
3
4
5
6
7
8
Tasche für Masch.Pistole
Masch.Pistole
R.V.3
14984 H.

piercing shells with high explosive filler) projectiles fired by these 10 cm and 12.8 cm guns was more than adequate to penetrate the armor of any enemy tank out to a range of 2000 meters, as shown in the following table:

	10 cm K. Pz.Sfl.IVa 10 cm Pzgr.rot	12.8 cm K.40 (Pz.Sfl.) 12.8 cm Pzgr.
Shell Weight	15.6 kg	26.4 kg
Vo	822 m/s	880 m/s
Range	Penetrates at 30 degrees	
100 m	173 mm	200 mm
500 m	155 mm	175 mm
1000 m	138 mm	150 mm
1500 m	124 mm	132 mm
2000 m	111 mm	130 mm

The combat capabilities of these heavy self-propelled guns is related in the following account written by the unit commander, Oberleutnant Kurt Hildebrandt:

*Finally it appears that what we all have waited for will occur, after retreating for days the enemy has finally positioned themselves for battle and this time our company is also there. Both of the other companies in our **Abteilung** have had more luck than us. Since the start of July, when the new offensive began, they had already encountered enemy tanks several times and knocked out 15 tanks within these three weeks.*

*This increases our burning desire to show that we understand how to go about our business with our three heavy guns, one **10.5 cm** and two **12.8 cm Sf. (Selbstfahrlafette)**. Because of engine problems "Max" (one of the **12.8 cm Sf.**) is not present, but we can still get the job done with "Moritz" and "Brummbaer". Also we possess an imposing firepower with our **4.7 cm Sf.** for use against light tanks and infantry targets.*

At dawn we advanced up to a ridge line without encountering significant resistance. The terrain falls gradually away from us and begins to rise again at a distance of 2 to 3 kilometers. The opposite ridge line is about 4 kilometers away.

The last morning fog disperses so that we can now more exactly orient our position. Nothing is stirring on the opposite side of the flat valley. No gunfire is heard.

But wait, what are the small protrusions along the opposite ridge line? Are they natural mounds or could they possibly be bunkers? We still can't make them out, because the rising sun is blinding us.

Then a gun commander yells: "Over there stands a tank" and pointed toward one of the small mounds. When sighted, the silhouette of a tank turret was recognizable. Soon after, the other "mounds" reveal themselves to be tank turrets, but nothing more is showing than the turrets with the clearly recognizable guns. We count about 30 tanks, located in a wide open curve at our elevation.

It will soon be revealed whether the enemy has noticed that he has been spotted and his hiding game is no longer useful. Maybe he has also spotted our movement, in any case, his tanks are firing at us. However, the range is too long and the shots are inaccurate and harmless.

The crews of "Moritz" and "Brummbaer", indeed

the entire combat echelon of the **Kompanie**, are understandably excited. One can only describe it as **Jagdfieber** (hunting fever). Still we must wait because it is not the task of **Panzerjaeger-Sf.** to charge in tank attacks. That is the role of our **Sturmgeschuetz** and Panzer.

After waiting for about three hours, a single tank drives at high speed into the valley toward us. Before "Moritz" can be pushed forward into a favorable firing position, the Red is already down in the valley. We can't engage this target without driving onto the front slope and revealing our position. From the engine noise and clattering tracks we surmise that he is now in running along the valley parallel to our ridge line. It is totally obvious that he will discover us.

A quick view over the terrain showed the company commander that the enemy tank must drive out of cover at a specific point if he keeps going in the same direction. "Moritz" was brought into a suitable position, but the Reds appear to have become suspicious. The track clatter is no longer heard, but the engine still rumbles.

Then the engine howls; we hear him advance taking a course toward a good position for spotting him. Now he appears, it's a T.34 which slowly drives along the depression with the turret turned toward us and continuously firing in our direction.

At this moment, "Moritz" fires the first shot at the Soviet tank. We follow the tracer, which hits about two meters in front of the crate. Apparently he didn't notice anything, because he didn't accelerate. It appears to us as unendingly long before the second shot is fired. The tank will shortly disappear before a small protrusion. Fire! This time the shot hit. Or maybe not? Have we fooled ourselves?

The tank rolls calmly along and we almost see him disappear, with only the turret showing.

Before we can voice our disappointment, the hatch opens and a Red falls more than jumps out and almost at the same time a spurt of flame bursts out of the turret followed by thick smoke. The tank burns. Also indeed a **Volltreffer** (direct hit)!

We hardly have time to celebrate our success when almost at the same time across the valley a second tank starts up and aims toward us. He isn't driving as fast and fires only with the machinegun. It's a KW.I. Now he halts, fires the gun two or three times, drives farther, and after halting a second time to fire, advances to within about 1500 meters.

"Moritz" had changed position, revealing the advantage of a **Sf.** in comparison with an emplaced **Pak** (anti-tank gun). The KW.I is in his sights. The KW.I spotted our gun and a duel begins between them. The Bolshevik fires several rounds in rapid succession that are well

Left and Below: As ordered on 15 May 1942, the Pz.Sfl. fuer 12.8 cm K.40 were issued to a Panzerjaeger-Zug assigned to Panzerjaeger-Abteilung (Sfl.) 521, and sent to the Eastern Front. (KHM)

aimed, but he is hit by our **Granate** (shell) before he can achieve a hit. We then see two of the crew get out and run toward the other ridge. "Moritz" pulls back into a small depression directly behind his firing position so that he can't be observed by the enemy.

In the interim, both men from the KW.I have got back over by their tanks, and not long afterward a T.34 drives rapidly into the valley. In a few minutes he is by the knocked out KW.I and halts directly in front of it. Men jump out of the T.34 and get busy by the KW.I. Before we discover that they want to tow away the damaged tank, both start to move slowly. But something isn't right, because they halt after a few minutes and renew activity on both tanks.

On orders from the company commander, "Moritz" pulls forward into the old firing position and fires **Sprenggranaten** (HE shells) at the Reds running around both tanks. The first shot is much too long, the second much nearer the target, but the third? What is that? The T.34 was hit on the rear and in a few seconds is in flames. A T.34 has been set on fire by a **Sprenggranate**, and that at a range of 1500 meters!

After a short hour, another T.34 is already by the immobilized KW.I. Again the crew attempts to tow it off. With a few shots, "Moritz" also hit this tank and set it on fire. All three **Kolosse** (giants) lie close together. Will the burning T.34s also set the KW.I on fire? We all think so, but the enemy must have the same thought, because what now occurs is hardly believable. Again a tank charges into the hollow and the crew jump to fasten the tow cable, mount up, and must at any moment start off.

As "Moritz" pulls into the old firing position on the ridge line, he is taken under heavy fire from the far ridge. They have finally spotted him, and at least ten tanks are aiming their guns at him. Most of the shots are off target; only one tank appears to have the range. But this time "Moritz" immediately hits the target with the first shot being a direct hit and the third T.34 that wanted to tow the KW.I burns brightly.

Now it is high time that "Moritz" pulls back. He can't turn, or else he will show his entire wide side. Also reverse gear. We all want somehow to help him move faster, and quake inside from the thought that he still could be hit. Only 10 meters left and he is safe. There, a hit directly in front of him and he is covered by a cloud of dust. Still 5 meters, still 3 meters, now he must completely disappear. At this moment, a round hits between the tracks under the hull and bores into the ground hardly one meter behind him. But, "Moritz" remains undamaged.

The enemy lost five tanks in their unsuccessful operation. Even for the Reds that is a bit much. He is quiet. Now and then a tank drives along the other ridge, but no longer venture into the valley.

Shortly after the noon meal, a thunderstorm broke over us that within several minutes turned the ground into impassable mud. All of the wheeled vehicles were stuck and even the tracked vehicles have difficulty moving forward.

The Soviets exploited this moment to attack with tanks and infantry. Hidden by the thunderstorm, they advanced to within 800 meters. "Moritz" and "Brummbaer" pull into favorable firing positions and take both of the frontmost tanks under fire. This time it only takes a few minutes until both T.34 are burning. A third turns to flee, followed by infantry. It's hit, tearing off the turret.

Shortly before nightfall, the company commander gives the order to leager. The **Sf.** are pulled together in a small ring, with the ammunition trucks and motorcycles in the middle. Guards are posted and begin their first round.

Then we hear several shots fired to the right. We know that the **1.Kompanie** is located there and in the interim has been pulled forward. Soon we also see where the shots were aimed. An enemy tank at about 3000 meter range is rapidly driving toward us. Tracers follow him; ricochets hiss into the air.

"Brummbaer" was again prepared for action and traverses the gun toward the target. But this time he doesn't need to advance. One shot hit the rear of the tank; it's on fire!

What we now see, we will never forget. Without slowing, the burning tank races along, like a giant torch that shines against the dark evening sky. It drives farther and farther. Haven't the crew noticed that their tank is on fire? He is within 1000 meters before it stops. One man climbs out of the driver's hatch; a second follows with difficulty. Neither of the other crew members can save themselves. For a long time, the tank burned through the night, shaken by one explosion after another.

We all unwind from the day's excitement. Everyone slept soundly, even the "**Kaffeemuehle**" (coffee grinder) that visits every night and showers us with blessings can't disturb this.

The guards hear tank noises only once, shortly before midnight. But this was a long way off, and the company didn't need to be awakened. The next morning the KW.I is missing that the three T.34 wanted to tow. Indeed, the Soviets recovered him at night.

The next day passed calmly. We saw individual tanks on the far ridge, but no shots were fired, and none came over to visit.

Since dawn our **Aufklaerer** (spotter plane) circles over us. About 0600 he dove down, pulled two circles, waved, and threw out a message. It fell near us and was taken to the company commander. This report, which was confirmed a half hour later by radio, read: "40 enemy tanks advancing from the northeast". Soon we receive a second message that an enemy tank assembly area has been spotted to the southeast.

*Rapidly, our **2.Kompanie**, two **8.8 cm Flak-Batterie** and a **Panzer-Kompanie** were pulled forward and held ready to our left. We've established on a front of about five kilometers. A deep curve about in the middle of our main battle line creates a so-called "**Schlauch**" (hose).*

*At night we again lager. Again the "**Kaffeemuehle**" circled above us. It did us good that one flying with its position lights on was shot down near us.*

*About 0200 in the morning, shortly before daybreak, the company pulled out of the lager. Hardly had the **Sf.** taken up position, when we heard tracks clatter and tank engines howling. Alarm! Our eyes tried to penetrate the early morning fog. There at about 1500 meters range the silhouette of a tank appears, there a second one, and there already another. The steel giants are charging out of the fog, firing from all guns.*

Still we haven't fired a single round; we're letting the Soviets enter the trap. "Moritz" and "Brummbaer" are located where they can hit the tanks in the flank. Suddenly the first shot is fired, followed by rapid fire from all the anti-tank guns. In about five minutes, 14 Soviet tanks are on fire with the flames thrown toward the red morning sky. A fantastic, unforgettable sight.

*That's too much for the enemy tanks; they turn away and disappear into the fog. Only two have broken through our lines and were destroyed as they advanced toward a **Flak-Batterie**.*

*A few minutes after the first attack, we heard wild firing to our right where our **1.Kompanie** was positioned. Soon came the call "Enemy tanks from the right!" "Moritz" and "Brummbaer" traverse their guns and knock out four tanks. It is lucky that the tanks left and right of us didn't attack at the same time but instead at an interval of about 15 minutes - or the situation could have turned bad.*

Still one more time during the morning, the opponent tanks tried to break-through to our right. Again he had to turn away because of losses, and again three tanks retreating past our assembly area didn't see us.

"Moritz" immediately grasped the situation and in a short time knocked out all three of these T.34.

*Forty-nine Soviet tanks, almost all T.34, were knocked out in a few hours this morning. Our **Abteilung** can claim 29 of these. Our **Kompanie** has passed its test in fire. In two days in combat they had knocked out 26 T.34 without a single loss ourselves.*

Panzerjaeger-Abteilung (Sfl.) 521 continued to advance with the Summer offensive to the east, reporting that both **12.8 cm Pak Sfl.** were still operational on 2 and 12 November 1942 (along with three **4.7 cm Pak Sfl.** and seven **7.62 cm Pak Sfl.**). One **12.8 cm Sfl.** was reported to be operational (along with three **4.7 cm Sfl.** and three **7.62 cm Sfl.**) with **Panzerjaeger-Verband von Langenthal** under A.O.K.6 in Stalingrad on 1 and 2 December 1942. Later, Henschel reported that both **Pz.Sfl.V** had been lost at Stalingrad.

Above: Nicknamed "Max" and "Moritz", both Pz.Sfl. fuer 12.8 cm K.40 survived to reach Stalingrad and were both reported to be operational directly before the Russian envelopment. (KHM)

Pz.Sfl. fuer 12.8 cm K40
previously known as Pz.Sfl.V

Weapons Data: 12.8 cm K.40 (L/71)
- Elevation: -15, +10 degrees
- Traverse: 12 degrees (7 R & 5 L)
- Gun Sight: Sfl.Z.F./Rdbl.F.
- Graduated to: 3000 m Pzgr., 3000 m Sprgr.

Secondary: 2 - 9 mm M.P.

Ammunition:
- 15 - 12.8 cm Pzgr. & Sprgr.
- 384 - 9 mm Patr.f.MP

Crew:
- Commander, Gunner
- Loader
- Driver, Radio Operator

Communication: Fu.Spr.Ger."a" & Intercom

Measurements:
- Length, overall: 9.70 m
- Length, w/o gun: 7.00 m
- Width, overall: 3.15 m
- Height, overall: 2.75 m
- Firing Height: 2.26 m
- Wheel Base: 2.67 m
- Track Contact: 4.75 m
- Combat Loaded: 36.5 metric ton
- Fuel Capacity: 450 liters

Armor Protection:
- Front: 50 mm
- Sides: 30/20 mm
- Rear: 15 mm
- Deck: 30/20 mm

Automotive Capabilities:
- Maximum Speed: 25 km/hr
- Avg. Road Speed: 20 km/hr
- Cross Country:
- Range on Road: 170 km
- Cross Country: 80 km
- Grade: 30 degrees
- Trench Crossing: 4.5 m
- Step: 0.8 m
- Fording Depth: 1.05 m
- Ground Clearance: 45 cm
- Ground Pressure: 0. kg/cm^2
- Power Ratio: 8.5 HP/ton
- Steering Ratio: 1.78

Automotive Components:
- Motor: Maybach HL 116 S
  - 6 cyl., water-cooled
  - 11.6 liter
  - 310 HP
- Transmission: Zahnradfabrik SSG 77
  - 6 forward, 1 reverse
- Steering: Differential
- Drive: Front sprocket
- Roadwheels: 8x2 per side
- Tires: Rubber 700/98-550
- Suspension: Torsion bars
- Track: Dry pin
  - Kgs 520/160
- Links per side: 85

Left: This Pz.Sfl. fuer 12.8 cm K.40 at a training base in Germany early in 1942 still has the fake sheet-metal compartment (being used as a seat) and a canvas cover protecting the open superstructure from the elements. (KHM)